CLASSIC DESIGN

设计典范 1

《设计典范》编辑组 编　鄢 格 译

辽宁科学技术出版社

图书在版编目（CIP）数据

设计典范/《设计典范》编辑组编：鄢格译．——
沈阳：辽宁科学技术出版社，2010.9
ISBN 978-7-5381-6065-9

Ⅰ．①设… Ⅱ．①设… ②鄢… Ⅲ．①商业－服务建
筑－空间设计－图集 Ⅳ．①TU247-64

中国版本图书馆CIP数据核字（2010）第129276号

出版发行：辽宁科学技术出版社
　　　　　（地址：沈阳市和平区十一纬路29号　邮编：110003）
印 刷 者：利丰雅高印刷（深圳）有限公司
经 销 者：各地新华书店
幅面尺寸：889mm×1194mm
印　　张：40
字　　数：80千字
附　　件：4
印　　数：1~3000
出版时间：2010年9月第1版
印刷时间：2010年9月第1次印刷
责任编辑：陈慈良
封面设计：曹　珍
版式设计：曹　珍
责任校对：周　文

书　　号：ISBN 978-7-5381-6065-9
定　　价：518.00元

联系电话：024-23284360
邮购热线：024-23284502
E-mail: lnkjc@126.com
http://www.lnkj.com.cn
本书网址：www.lnkj.cn/uri.sh/6065

Contents 目录

OFFICE 办公

006. Paul, Hastings, Janofsky & Walker, LLP – Paris / 美国普衡律师事务所巴黎办事处
012. Paul, Hastings, Janofsky & Walker, LLP – Los Angeles / 美国普衡律师事务所
018. Manchester Square / 曼彻斯特广场
024. Jamba Juice Support Centre / 嘉宝果汁办公中心
030. Gummo Advertising Agency / Gummo广告公司
036. Larchmont Office / 拉奇蒙特办公
042. New Google Meeting Room / 谷歌的新会议室
046. Baroda Ventures / 巴洛达办公
052. Office Herengracht 433, All Capital / 全资投资集团，赫恩格拉希特433号办公室
058. Saatchi + Saatchi LA / 萨奇广告公司
064. Office Lab / Lab办公室
068. Turner Duckworth Offices / 特纳·达克沃斯平面设计公司
072. The Information Box / 信息盒子
076. FOX Latin America Channel Base Offices / 福克斯北美频道办公
082. China Everbright Limited (CEL) / 中国光大股份有限公司
088. Ernst & Young University / 安永大学
094. An International Firm – Shenzhen Branch / 一家跨国公司的深圳分公司
100. Chengdu Sales Office / 成都销售处
106. Machiya Office / 町屋办公室
110. City Year Headquarters for Idealism / "为理想奋斗"城市年鉴总部
116. Çağdaş Holding Office / 卡达斯公司
120. Meinl Bank / 麦因尔银行
124. DITTEL – Architekten / 迪泰尔建筑工作室
128. Headquarters Expansion / 诺和诺德制药公司北美总部
134. The LG Air-conditioner Academy / LG空调办公大厦
138. Ding Pu Hi-tech Square / 嵿埔之星科技广场
144. TIC / 东京旅游信息中心

RESTAURANT 餐厅

148. Shiyuan restaurant / 食源餐厅
154. Maedaya Bar / 马依达雅酒吧
158. Adour Alain Ducasse / 阿杜尔·艾伦·杜卡斯餐厅
164. Adour at The St. Regis Washington, D.C. / 华盛顿圣瑞吉酒店的阿杜尔餐厅
168. Dos Caminos / 多斯·卡米诺斯餐厅
174. Matsuhisa / 马苏希萨餐厅
180. Wildwood Barbeque / 怀尔德伍德烧烤屋
186. Tiandi Yijia – Restaurant / 天地一家餐厅
192. BEI – Asian Restaurant / "北"—亚式餐厅
198. Sureno (Mediterranean Restaurant) / Sureno—地中海餐厅
204. Cityscape Restaurant / 城市景观餐厅
210. NEVY / 聂维餐厅
216. Vengeplus Capacity / 闻佳餐厅
222. SEVVA Restaurant / SEVVA餐厅
228. BRAND Steakhouse / 品牌牛排餐厅
232. Inamo / 埃那谟餐吧
236. "pearls & caviar" / "珍珠和鱼子酱"
242. Lido / 莱多餐厅
248. VLET Restaurant / 韦丽特餐厅
252. Chocolate Soup Café / 巧克力汤餐厅

Contents

256. Mc Donalds Urban Living Prototype / 麦当劳—城市生活设计原型
260. Oth Sombath Restaurant / 奥斯餐厅
266. Restaurant Alain Ducasse / 艾伦—杜卡斯餐厅
270. Leggenda Ice Cream and Yogurt /莱真达冰激凌乳酪店
274. Rosso Restaurant / 罗索餐厅
278. VYTA – Boulangerie / 维塔面包店
282. EL JAPONEZ / EL JAPONEZ餐厅
286. TCharcoal BBQ 3692 / 3692烤肉餐厅
292. Le Square Restaurant / 广场餐厅
298. Sky 21 / 天空21号

HOTEL 酒店

302. Kush 222 / 库什222号
306. Chambers MN / 明尼酒店
312. Carbon Hotel / 凯博恩酒店
318. The Dominican Hotel / 克劳斯K酒店
324. The Dominican Hotel / 多米尼加酒店
330. Jeronimos Hotel / 杰尼贸酒店
336. The Levante Parliament / 累范特风酒店
342. CitizenM Hotel / 旅行者酒店
348. Murmuri Hotel / 莫姆里酒店
354. Hotel Sezz / 瑟兹宾馆
358. The George Hotel / 乔治酒店
364. Villa Florence Hotel / 佛罗伦萨别墅饭店
370. JW Marriott Hotel Hong Kong / 香港JW万豪酒店
374. Hotel Pirámides Narvarte / 那瓦特金字塔酒店
378. Mövenpick Airport Hotel / 慕温匹克机场酒店
382. The Library /图书馆大酒店
388. X2 Koh Samui /苏梅岛X2酒店

CULTURE 文化

394. Thurles Arts Centre and Library / 瑟勒斯艺术中心及图书馆
400. A.E. Smith High School Library / 阿尔弗德·E·史密斯高中图书馆
404. Wagner Middle School Library / 瓦格纳中学图书馆
408. Ps 11r Primary School Library / Ps 11r小学图书馆
414. The Danish Jewish Museum / 丹麦犹太人博物馆
418. Museum / 博物馆
424. New Acropolis Museum / 新卫城博物馆
428. The Art Institute of Chicago – The Modern Wing, Chicago / "现代之翼"博物馆
432. The Arts of Asia Gallery, Auckland / 亚洲艺术画廊
436. That's Opera / "那是"剧院
442. Bachhaus Eisenach / 巴赫纪念馆
446. Multikino Szczecinl / 玛尔缇凯诺什切青电影院
450. Blue Room Theatre at Chesapeake / 宝龙影院
454. Ordrup School / 奥竹普学校
460. Medical Faculty, University of Groningen / 格罗宁根大学医学部

FITNESS CENTRE 健身休闲

464. Chesapeake Fitness Centre, East Addition / 切萨皮克健身中心扩建
468. 14 Street Y Renovation / 14号街

472. Y+ Yoga Centre / Y+瑜伽中心

EXHIBITION 展示

476. Breaking New Ground / 新天地

480. Butterflies + Plants / 蝴蝶与植物展厅

484. Fuji Xerox Epicenters / 富士施乐中心

488. Stylecraft Showroom / 风尚家居展示厅

492. Bernhardt Design Showroom / 伯尔尼设计公司展厅

496. BRANCH in Changchun / 长春藤

500. FlatFlat in Harajuku / 原宿平平游戏体验店

504. homeculture / 文化屋

508. GRAFTWORLD – Exhibition in the Aedes Gallery / Graft作品展厅

512. Moonraker / 蒙莱克尔

518. Internorm Flagshipstore / Internorm办公展厅

522. Heaven / 悦天展示

STORE 商业

526. Tesla Store / 特斯拉汽车专卖店

530. ProNature / 天然坊

534. Healthy Life Natural Health Food Store / 健康生活天然保健品店

538. Barbie Shanghai / 上海芭比娃娃旗舰店

544. WEEKDAY MALMÖ / 周末假日时装店

550. Monki Stockholm / 曼奇品牌店

554. MONKI 1: Forgotten Forest / MONKI品牌零售店–遗忘的森林

558. LEVI'S BB Barcelona Summer / 李维斯

562. Forum Duisburg / 杜伊斯堡广场购物中心

566. Sexta Avenida / 第六大街购物中心

570. Celebrity Solstice / 名至巡游艇

574. Gourmet / 美食购物广场

578. Only Glass / 玻璃

582. Rocawear Mobile RocPopShop / Rocawear移动服饰店

586. SENSORA / 森索拉

590. UHA Mikakuto / 味觉糖食品店

594. O2 / O2通讯

598. Lefel – The Whispering World / 拉菲服装店–幽静的世界

602. Nike Airmax 360 Exhibition / 耐克

606. Kymyka shoes and bags, Maastricht / Kymyka鞋包店

610. Offspring Camden / Offspring品牌鞋店

614. Sabateria Sant Josep / 桑特·何塞普鞋店

618. The Eyecare Company / 眼睛保健眼镜店

624. FeiLiu Fine Jewellery / FeiLiu精品珠宝店

628. Couronne / 钟表珠宝店

632. VAID Ginza / VAID珠宝专卖店

638. INDEX / 索引

Paul, Hastings, Janofsky & Walker, LLP—Paris

美国普衡律师事务所巴黎办事处

Project Name: Paul, Hastings, Janofsky & Walker, LLP—Paris **Designer:** Progetto CMR - Massimo Roj Architects **Location:** Paris, France **Photographer:** Eric Laignel **Time:** September 2007 **Total Size:** 48,000 sf

项目名称：美国普衡律师事务所巴黎办事处 设计师：波接特－马西莫 罗伊设计所 项目地点：法国，巴黎
摄影：埃里克·莱涅尔 完成时间：2007年9月 面积：4460平方米

Unique to this project, walls could not be reconfigured in the historic building, so design was focused on furniture, cabinetry and the creative use of light. Careful planning allowed for functional segregation of public reception rooms from private practice areas while respecting the historic fabric of the building. The furnishings are an eclectic combination of classic pieces with sleek custom conference tables and seating unified by a restrained palette of cream and dove gray upholstery. The color scheme includes a subtle collection of grays and beige that build upon the historic features of the building with a diverse range of texture and sheen to create visible depth. There are gilded and gold-painted walls combined with back-painted glass top tables. An endlessly spiraling staircase rises from the sunbathed lobby with built in banquette and an up-lit dome over the reception desk. The lighting design is a sophisticated blend of restored original fixtures, bespoke Murano chandeliers, concealed cove up-lighting and contemporary backlit glass elements.

The project, which included approximately 4460 square metres spread among eight floors, was designed to accommodate 134 attorneys and staff, a conference floor and prominent library. The primary materials used were limestone, artisanal plaster, limed European oak, wrought iron and glass.

这个项目特殊的地方是历史性建筑的墙壁不能拆除，因此家具和灯光成为设计的重点。设计师经过仔细规划，在尊重历史建筑结构的同时，将公共接待室与私人办公区分割开来。家具选用经典的款式，特别定制的豪华桌椅采用乳白色和灰色搭配。整个空间以灰色系和米色搭配，以反映建筑的历史风貌，并用带有不同纹理和光泽的材料营造深度感。镀金墙壁上镶着单面上漆的玻璃。大厅的走廊里一条旋转楼梯盘旋而上，接待处上方的圆屋顶让整个大厅都沐浴在阳光里。灯光设计十分复杂，包括对原有的装置进行修复，安装定制的马伦诺吊灯，隐藏式穹形顶灯和现代的玻璃背光灯。

办事处的面积约为4460平方米，有八个楼层，要容纳134位律师及员工，还有会议楼层和最为重要的图书馆。材料主要是石灰石、石膏板、涂刷过石灰水的欧洲橡木、熟铁和玻璃。

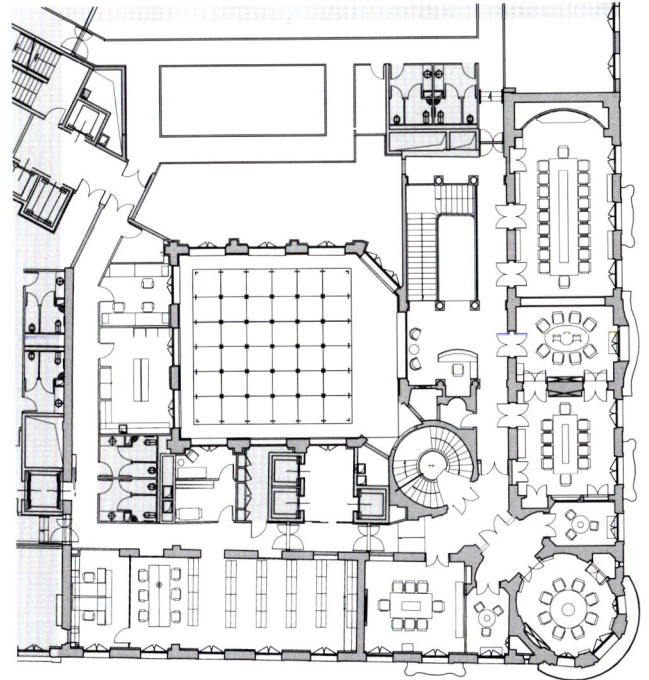

Paul,Hastings,Janofsky & Walker,LLP – Los Angeles

美国普衡律师事务所洛杉矶总部

Project Name: Paul, Hastings, Janofsky & Walker, LLP – Los Angeles
Designer: Progetto CMR - Massimo Roj Architects **Location:** Los Angeles,USA
Photographer: Benny Chan   **Time:** 2007   **Total Size:** 194,400 sf
项目名称：美国普衡律师事务所洛杉矶总部 设计师：波接特－马西莫 罗伊设计所 项目地点：美国，洛杉矶 摄影：班尼·陈
完成时间：2007年 面积：18060平方米

Paul, Hastings, Janofsky & Walker, a multi-disciplinary law firm with offices worldwide, sought a unifying design scheme for its offices that reinforces the firm's corporate identity and branding efforts and reflects the culture of one of the nation's most dynamic law firms. Each office employs the newly designed standards, yet is able to maintain the distinct personality of the individual offices.

Paul Hastings' Los Angeles headquarters were designed to endorse the culture established by the law firm's young, vibrant managing partners. The partners sought a design that would reflect their commitment to an expanding enterprise, innovative solutions, and a progressive environment.

The office environment was designed with a modern, neutral palette and angled planes of drywall that reflect and enhance natural light throughout the 190,000 square foot space. The main reception area exemplifies the design approach — the two-storey atrium space is filled with natural light and colour, providing a perfect backdrop for dramatic views of Los Angeles. The attorney offices were designed with custom, opaque glass walls and clerestories to draw natural light deeper into the space. The ethereal quality of the office design is calming and epitomizes Paul Hastings' business approach; yet, the richly textured design materials and careful detailing offer a counterpoint to the lightness of the space and reinforce the law firm's strength and depth.

普衡律师事务所是一家综合的律师事务所，在全球各地都设有办事处。事务所要求其各个办事处的装修风格统一化，以加强企业形象和品牌知名度，反映其打造最具活力的律师事务所的企业文化。每个办事处都采用了新的设计标准，拥有独特的个性。

普衡律师事务所洛杉矶总部的设计理念以事务所年轻、充满活力的管理团队所创立的企业文化为基础。管理团队希望事务所的设计，能够反映出他们一直所坚信的信念，即扩大公司规模、开拓创新思维和打造一个积极向上的环境。

办公环境的设计选用现代的中性色调，不同角度的墙面有助于为这个18060平方米的空间反射和增加自然光。主接待区是设计技巧的集中展现——这里有一个挑空的中庭，自然光充足，色彩丰富，还有美妙的洛杉矶景色供客人欣赏。律师办公室依照风俗习惯量身打造，采用不透明的玻璃幕墙和高悬的天窗让自然光线深入空间之中。办公室的设计精致稳重，是普衡经营理念的缩影。然而，质感丰富的装饰材料和精心设计的细节活跃了空间的气氛，并强调了事务所的实力和品质。

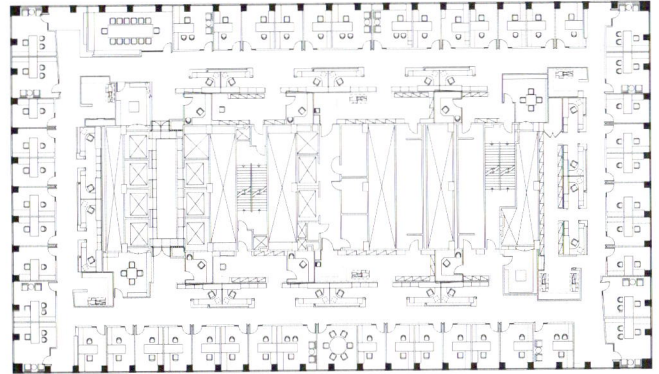

Manchester Square

曼彻斯特广场

Project Name: Manchester Square   **Designer:** SHH   **Location:** London, UK   **Photographer:** James Silverman   **Time:** 2008

项目名称：曼彻斯特广场 设计师：SHH事务所 项目地点：英国，伦敦 摄影：詹姆士·西尔弗曼 完成时间：2008年

It is a new London office interiors scheme for a private company in London's west end, designed by SHH, working closely with client-side design director Zeljko Popovic. The Manchester Square offices are the result of a brief to create a "high impact, 21st century office interior with a strong personality" with more in common with a gentlemen's club than a traditional office space. The requisite air of subtlety and elegance in the scheme is reflected in the complete lack of external branding at the company's front door.

Although the property was made up of a classic (and Grade II-listed) Georgian townhouse with a former stables and garage mews building to the rear, the client was very open to a sense of contrast for the scheme, favouring a highly contemporary treatment. The inherited interior — also previously an office — was not particularly grand, but any standout details were to be retained (such as marble mosaic floor tiles in the entrance area, a number of marble-inlaid fireplaces and highly detailed ceiling cornicing).

The building has five storeys and the rear mews has two storeys, with the upper ground mews floor corresponding to the ground floor of the main space. The original connection between the two was via an open courtyard, which SHH has now partly built over to form a covered walkway between the two for all-weather access.

本项目位于伦敦西区，是SHH设计公司与委托方设计代表利科·波波维奇为一家私人公司共同设计的一间办公室。"曼彻斯特广场"办公室基于这样的理念：创造一个有高度影响力的21世纪室内办公环境，它要拥有强烈的个性化特点，不拘泥于传统办公空间，使之看起来像是绅士俱乐部一样！本项目理念中的优雅和浪漫迷离状态直接体现在公司的前门：没有把公司的商标体现在外面！

虽然本项目原是乔治亚州一栋传统的乡镇建筑（后部有传统的马厩和车库），但业主的想法却是：把它建成一间高度现代化的办公室！建筑原来的内部空间也是办公环境，虽然不是很壮观，但是设计师们还是保留了一部分原来的样子（例如：入口处的大理石马赛克地板、用大理石镶嵌的壁炉以及棚顶精细的檐口）。

本建筑有五层楼，建筑后部的车库有两层楼，车库的顶层地板与主空间一样高。起初，这两栋楼是通过一个开放的庭院连接的，如今SHH设计公司把庭院的部分建成了一条适应各种气候的通道。

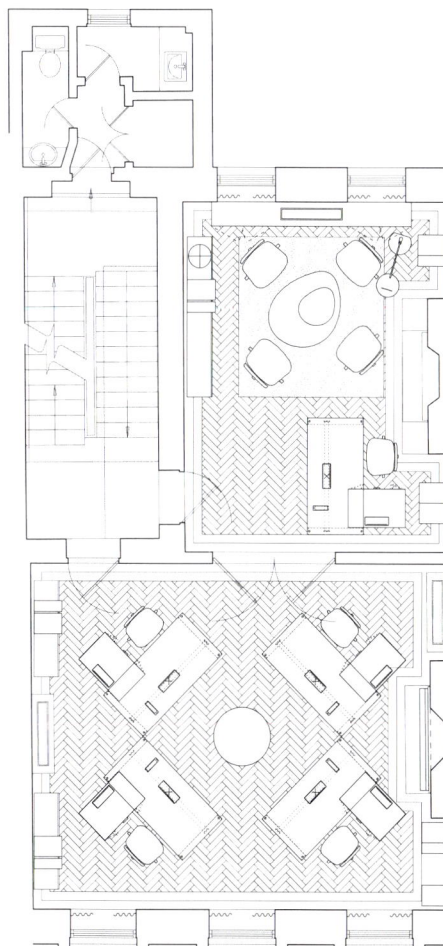

Jamba Juice Support Centre

嘉宝果汁办公中心

Project Name: Jamba Juice Support Center   **Designer:** Pollack Architecture
Location: Emeryville, CA, USA   **Photographer:** Cesar Rubio; Eric Laignel   **Time:** 2008
Area Size: 3344 sqm

项目名称：嘉宝果汁办公中心 设计师：波拉克建筑事务所 项目地点：美国，加州，埃默里维尔 摄影：塞萨尔·鲁比奥
埃里克·莱格尼尔 完成时间：2008年 面积：3344平方米

California-based Pollack Architecture was hired by Jamba Juice to transform a Class B office building of steel, glass, and veneer brick into an energetic office space that captures the brand's spirit while paying homage to the funky warehouse from which the national smoothie chain was moving. The architects renewed an existing building to reflect the dynamic culture of the company while creating an industrial/warehouse quality for the space. Signature shades of vibrant orange and lime green are featured throughout the office space, as are numerous graphics relating to the Jamba Juice brand. Doors throughout the space have been labeled with brand identifying names, such as Blender, Carrot, Banana, and Strawberry. The design connects the company's health-centric philosophy with key elements of its retail environment to create a fun and energizing workplace.

The spacious reception area features a signature "Jamba-green" wall that holds a branded image and welcomes visitors to both the office space as well as the brand. A near floor-to-ceiling steel industrial window wall system, typically used in warehouses, separates the reception from the conference and board rooms, while allowing natural light to penetrate throughout the spaces. Overhead, air ducts and structural beams remain exposed to immediately reveal warehouse characteristics for an industrial look.

The transition to the upper level includes a staircase, framed with a stainless-steel, glass, and wood guardrail, along a "Jamba-orange" wall. Individual workstations are located on this floor along with small-group meeting rooms.

波拉克建筑事务所受邀负责嘉宝果汁办公中心的改造——将一幢钢筋玻璃打造的二等办公大楼改造成活力充沛、彰显品牌精神的办公中心。设计师充分运用橙色和绿色，增添空间动感的同时又不失保留古旧的仓库特色。空间内所有的门上都贴上了不同水果图案构成的品牌标识，彰显公司一直秉承的以健康为中心的理念。

接待区域内，一面装饰着品牌特有的绿色的墙上凸显了嘉宝特有的形象，同时指引着来访者达到办公空间。几乎与屋顶等高的窗户幕墙结构营造了仓库特有的感觉，将接待处同会议室分隔开来，同时将自然光线引入。头顶上，排气管道和梁结构直接裸露在外。

不锈钢及玻璃材质打造的楼梯带有木质栏杆，沿着墙壁向上一直通往楼上的办公区。独立的工作台与小型会议室平行设置。

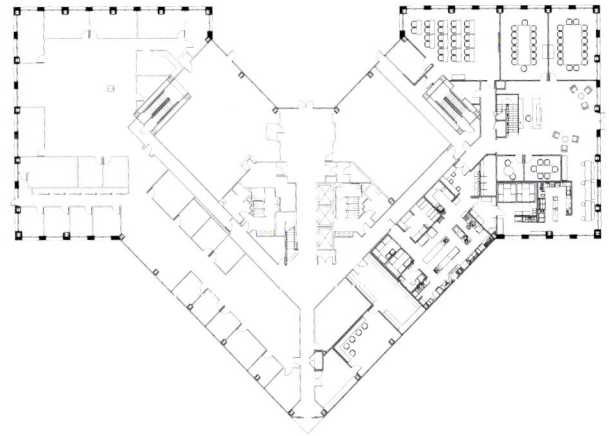

Gummo Advertising Agency

Gummo广告公司

Location: Amsterdam, Netherlands Designer: i29 interior architects Photographer: i29 interior architects Completion date: 2009

项目地点：荷兰，阿姆斯特丹 设计师：i29室内设计 摄影师：i29室内设计 完成时间：2009

As Gummo were only going to be renting the space on the first floor of the old Parool newspaper building in Amsterdam for two years, i29 convinced Gummo to embrace the mantra of "reduce, reuse, recycle" to create a stylish office space that would impact as little as possible on the environment or their wallets. They developed a theme that reflects Gummo's personality and design philosophy — simple, uncomplicated, no-nonsense, yet unquestionably stylish with a twist of humour. Everything in the office conforms to the new house style of white and grey. All the furniture was locally sourced via Marktplaats (the Dutch eBay), charity shops and whatever was left over at the old office. Everything was then spray painted with polyurea Hotspray (an environmentally friendly paint) to conform with the new colour scheme. Even Jesus wasn't immune, as you can see in the attached pictures. The new office is a perfect case study of a smart way to fill a temporary space stylishly and at minimal cost. The collection of old and repaired products in its new coating has given a new potential and soul to the old furniture.

因为Gummo广告公司只租用Parool报业大楼两年，所以i29说服他们采用"降低成本、循环利用"的理念，打造一个时尚的办公空间，既节约预算又不破坏环境。他们在设计中加入了Gummo广告公司的个性设计理念——简单、实际、不复杂，同时又兼具时尚感。每间办公室都采用了简约的白色和灰色风格。所有的家具都来自网上购物、慈善商店、以及从前遗留下来的旧家具。为了美观和色彩统一，所有家具都采用了Hotspray喷漆（一种环保漆）。在如何让一个临时的空间时尚又简约方面，这个项目是一个典范。为旧东西重新包装喷漆就像是赋予了他们全新的潜力和灵魂。

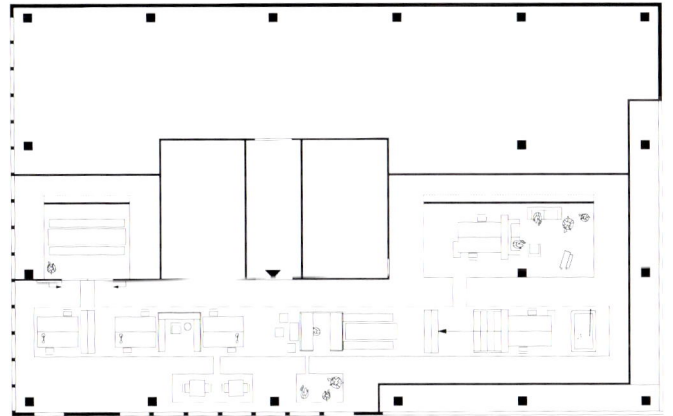

ONLINE
MONEY
MEDIA
PUBLISHER'S
BIZON
HET PAROOL
VOLKSKRANT
JC DECAUX
BOOMERANG
ATS PR
RTV NH PR
NL20 PR
ADFORMATIE
NS
GVB
RIETVELD
DESIGN ACADEMIE
WILLEM DE KONING
MEDIA COLLEGE
MIAMI AD SKOOL
FONTYS
HKU
HKA
ARTEZ
MAASTRICHT
MEMORABLE MOMENTEN
SHOE
J KRAMER
DRIETWEE
EXHIB. SPACE
GEMEENTE
Twitter
EMBEDDING
TM 4 KvZ

Larchmont Office

拉奇蒙特办公

Project Name: Larchmont Office   **Designer:** Rios Clementi Hale Studios   **Location:** Los Angeles, CA, USA   **Area Size:** 17,000 sf

项目名称：拉奇蒙特办公 设计师：当瑞斯.克里蒙蒂 哈勒工作室 项目地点：美国，加州，洛杉矶 面积：1579平方米

The architects transformed the building into a space of light by tearing down the exterior walls and installing a floor-to-ceiling window wall system, while maintaining the structure's existing footprint. The main entrance leads patrons to the receptionist counter. The white surfaces of the area allow light to travel deeper into the building through the front doors and around the space over lacquered finishes and bright colors. Next to the reception area is a large open conference room with 37-foot-high sliding glass doors. A fully-equipped kitchen is located around the corner to the west, featuring notNeutral products and designs. Around the corner to the east is a small powder room that showcases a floor-to-ceiling wall of green turf. Also located on this level is another informal meeting space, which the staff refer to as the "tree house" since the room features a floor-to-ceiling front window and overlooks a large tree. The space houses comfortable chaise loungers and a round wood table.

The upstairs offers two informal conference rooms surrounded by workstation pods. The goal was to create a nonhierarchical studio environment, so the architects designed two different pad scenarios where either six individuals are grouped together in the same area or the one large pod that incorporates 19 individual workstations.

Natural light is provided throughout the second floor with the surrounding glass-encased balconies, providing an indoor-outdoor atmosphere and offers additional gathering and break spaces. Light is also brought in by the existing and newly installed skylights. The exposed ducts and structural systems with the use of foil at the ceiling create an industrial feel to the space, bringing the firm's idea of the office space being its "idea factory" to life.

设计师们将建筑原有的外墙去掉,通过落地玻璃幕墙系统把建筑转化成一个光亮的空间。办公室的正门将顾客们引导到接待区,白色墙面经过光线的照射,使得空间更加开阔。旁边是一个宽敞的开放式会议室,安装着落地玻璃滑动门。房间里有两张长会议桌,白色桌面,周围环绕着木头和钢椅子。其西侧是一间厨房,放满notNeutral的产品和设计,东侧是一间小化妆室。通往二楼的楼梯墙壁被用来展示照片、过去的模型、公司当前和未来的项目等。二层是几间小会议室,其中一间被叫做"树屋",落地窗正对街道,窗下还有一棵大树,将绿意迎进了室内。

在办公空间设计中,目标是创造出一个不分等级的工作氛围。因此设计师创造了两种类型的工作环境,并且计划使员工们在这些空间内进行轮换,以创造新的混合。此外,这里还设有可容纳19人的大工作间,一排排的桌子按照传统开放式办公空间的布局放置在裸露的金属箔天花板之下。

自然光透过环绕二层的玻璃阳台入射进来,创造出室内室外相互交融氛围,同时提供了额外的聚集和休息空间。此外,光线也可通过天窗进入。裸露的管道结构以及天花板上的金属箔,营造出一种工业感觉,让人不禁回想起安迪·沃霍的工作室,诠释着"生活的理念工厂"这一主题。

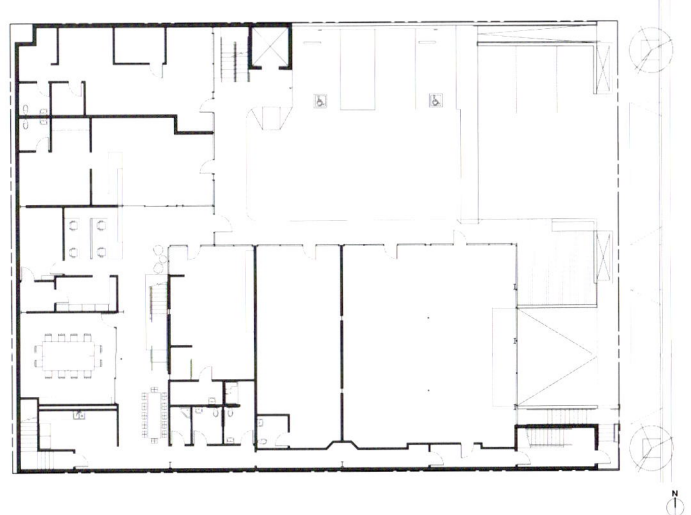

New Google Meeting Room

谷歌公司新会议室

Project Name: New Google Meeting Room   **Designer:** Camenzind Evolution Ltd.
Location: Switzerland   **Photographer:** Camenzind Evolution Ltd.   **Time:** 2008

项目名称：谷歌的新会议室 设计师：卡门辛德发展公司 项目地点：瑞士 摄影：卡门辛德发展公司 完成时间：2008年

A key element in the design approach was that the Zooglers should participate in the design process to create their own local identity. Under the guidance of the Director of International Real Estate at Google in Mountain View, an interactive and transparent approach to the architectural process was implemented from the beginning. A diverse team of local Zooglers were formed as the steering committee to represent the entire office.

Zooglers needed to be diverse and at the same time harmonious whilst making it a fun and an enjoyable place to work in. The survey also showed that while personal workspace needed to be functional and more neutral, communal areas had to offer strong visual and more aesthetically enjoyable and entertaining qualities to stimulate creativity, innovation and collaboration.

The Zooglers decided early on that they preferred to reduce their personal net area of workspace in order to gain more communal and meeting areas. The working areas were therefore designed with a high degree of space efficiency. Additionally, they had to be able to accommodate frequent staff rotation and growth. On average a Googler moves twice a year within the building, consequently the office layout was designed for maximum adaptability so that all groups and departments can use any part of the office space. The office areas are organized along a central core and are a mixture of open-plan workspaces for 6-10 people and enclosed offices for 4-6 people.

设计中的关键因素是请谷歌的员工参与设计过程，建立属于自己的特色。在谷歌国际房地产总监的指导下，一项透明互动的设计方式开始在山景城实施。员工们还派代表组成了设计工程的督导委员会。

设计师迅速地研究和分析了建筑中的优势和缺陷，以及员工们的实际需求，并对所有员工进行调查，开展了一系列的讲习班和演讨会，了解他们的需求。调查在心理学家的指导下展开，调查结果反映了不同员工的人格类型、代表阶层、价值观念和激励因素等信息。

调查结果是保密的，但这一过程表明，谷歌的员工需要多样化的工作环境，同时还要富于趣味性，令人心情愉快。调查显示，员工的工作场所强调功能性，个性化色彩不太强烈；而公共区域则要有强大的视觉冲击力，增加美的享受，还要有娱乐性，以促进创造、创新和协作。

员工们决定减少自己办公室的面积，以增加公共区和会议室的面积。因此办公室的设计充分利用了空间。另外，办公室还要能适应频繁人事变动和人员增长。谷歌员工平均一年要更换两次办公室，因此办公室布局的设计要具有强大的适应性，让各种人群和不同部门在这里工作起来得心应手。办公区设在各楼层的中心地带，开放式混合办公室能容纳6-10人，封闭式办公室能容纳4-6人。

Baroda Ventures

巴洛达办公

Project Name: Baroda Ventures   **Designer:** Rios Clementi Hale Studios   **Location:** Beverly Hills, CA, USA   **Time:** 2008   **Area Size:** 4,175 sf

项目名称：巴洛达办公 设计师：瑞斯 克里蒙蒂 哈勒工作室 项目地点：美国，加州，贝弗利希尔斯 完成时间：2008年 面积：388平方米

The architects adapted a surprising, yet delightful, combination of retro and contemporary styles for the renovation of the two-storey Baroda Ventures office and applied several themes throughout the design — classic modern furnishings with unexpected fabrics, elaborate ceiling medallions and doors escutcheons, glossy surfaces, and repeated patterning at various scales — while incorporating plentiful daylight. Timeless elegance was achieved by pairing traditional essentials with the latest in design to create a place of sophisticated opulence.

A custom-designed steel-and-glass staircase provides access to the second floor, affording uninterrupted views of the fountain below. At the top of the staircase, through a wooden door adorned with an exaggerated escutcheon, is the reception area, featuring a two-seat Eames Sofa in striped upholstery and an aqua leather Barcelona Chair. The ceiling was removed and seismically restructured to expose the existing hip roof. New skylights were installed to bring views of the sky and daylight deeper into the office. Interior walls are clad with either white back-painted glass, limed oak panels, or modern versions of tongue-and-groove, which function as textural foils for the existing masonry shell.

The nine private offices, surrounding the reception area and extending to the back, have been individually characterized while encompassing design elements found throughout the space. The architects contrasted exposed brick in each office with lacquered finishes and the use of glass. Used at the top three feet of interior walls and at the center of doors, abundant glass keeps the space open and well lit. An interior light well becomes the internal circulation core as a skylight allows natural light to travel throughout the space, bouncing off the white-painted glass walls of the stairwell.

设计师在这一翻修项目中试图运用多个主题——现代风格家具、精致天花图案、光亮的外观饰面，大小各异样式相同的图案占据整个空间，将现代与古典完美融合，营造了永恒经典的精致办公环境。

定制的不锈钢玻璃楼梯一直通往楼上，经过带有图案装饰的木门便可达到接待处。首先映入眼帘的是，点缀着条纹图案的双人埃姆斯沙发和浅绿色的皮质座椅。原有的天花板被拆除，取而代之的是全新的天窗，将美丽的蓝天景色和温暖的阳光引入进来。墙壁表面或是镶嵌着白色橡木板或是舌樵，为原有的石质结构增添了保护层。

接待区周围是9间单独的办公室，一直延伸到空间最后面。设计师在追随共性的同时又赋予其各自不同的特色，玻璃材质营造开阔通透感。采光井成为这一区域的中心结构，让光线源源不断地入射进来。

Office Herengracht 433, All Capital 全资投资集团,

赫恩格拉希特433号办公室

Project Name: Office Herengracht 433, All Capital   **Designer:** i29 interior architects / eckhardt&leeuwenstein architects   **Location:** Amsterdam, the Netherlands   **Photographer:** i29 interior architects   **Time:** October 2008   **Area Size:** 240 sqm

项目名称：全资投资集团，赫恩格拉希特433号办公室 设计师：i29室内设计事务所 / 伊克哈德&里温斯坦建筑师事务所
项目地点：荷兰，阿姆斯特丹 摄影：i29室内设计事务所 完成时间：2008年10月 面积：240m²

The board of All Capital, an investment group in capital stock, wanted to have a self-called "power office". i29 interior architects and Eckhardt&Leeuwenstein, two offices which collaborated during this project, created this by placing every board member in the spotlight on a playful way.

All three boardrooms and a lounge are executed in an overall design concept. Large round lampshades, spray painted gold on the inside, seem to cast light and shadow oval marks throughout the whole space. By this, a playful pattern of golden ovals contrasts with the angular cabinets and desks, which are executed in black stained ash wood. In the flooring the oval shaped forms continue by using light and dark grey carpet. Also, these ovals define the separate working areas.

The lounge area has, in combination with the white marble flooring, these same light/shadow patterns that cover the bar and benches in silver fabrics. This area can be used for presentations or social working, with an integrated flat screen in the bar and data connections in all pieces of furniture. The existing space is set in a 17th century historic building, at one of the most famous canals of Amsterdam called "de gouden bocht". All existing ornaments and details are painted white.

The keynotes for this company are money and power. The design concept expresses this by setting all members of the board literally in the spotlights. The golden and silver ovals shatter through the spaces like golden coins. Where it is all about, in investment and stock trading.

"全资"是主要致力于股本管理的一家投资集团，他们希望拥有一间自己所期待的那种"激发能力的办公室"。 i29室内设计事务所和伊克哈德&里温斯坦建筑师事务所合作完成了这个项目。设计中，通过将团队每位成员的座位以一种活泼的方式置于聚光灯下，达成了"全资"的这一夙愿 。

"全资"总部的全部空间包括三个会议室和一个休闲室，均秉承总体性的设计理念。 大大的圆形灯罩内壁用喷漆饰以金色，似乎能将或明或暗的椭圆形标记投射到整个空间。在这些灯具附近，黑色岑木制作的壁橱上设计了看起来饶有趣味的金色椭圆图案，与棱角分明的壁橱和桌子形成了美观的对比。地面则运用浅色和深灰色的地毯使这些椭圆图案得以延伸，同时这些椭圆还起到了划分不同工作区域的作用。

休闲室区域铺装了白色大理石地面，与会议室同样的明暗相间的椭圆图案出现在饰以银色织物的横梁和长凳上。这个区域用来举办报告讲座或者开展群体工作，并为此在横梁上安装了大屏幕，在所有家具上预留了数据接口。该办公室坐落在阿姆斯特丹著名运河城市之一、被称为"黄金岔路口"的一幢17世纪建成的历史悠久的建筑中，其所有装饰和细节都运用了白色。

这家公司最为重要的核心特征就是财富和力量。设计理念中通过将团队的所有成员逐个安排在聚光灯下来展示着这些特征。布满这个空间的金色和银色椭圆，恰如该集团在投资和股票交易中随处可见的金币一样令人过目难忘。

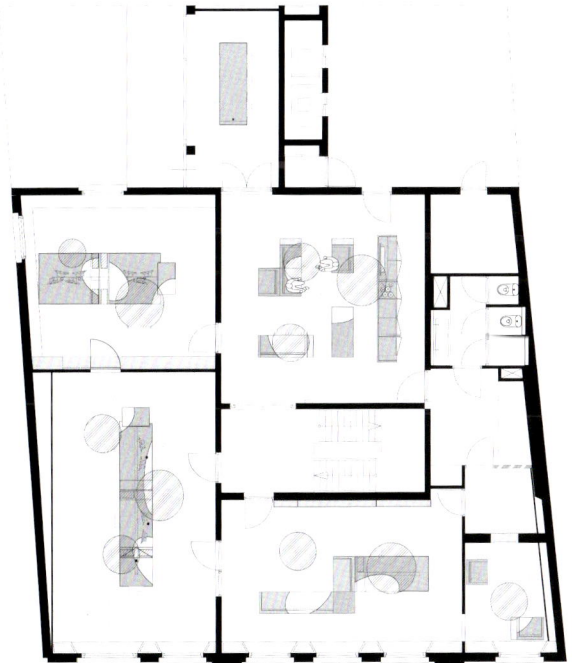

Saatchi + Saatchi LA

萨奇广告公司

Project Name: Saatchi + Saatchi LA   **Designer:** Shubin + Donaldson Architects
Location: Torrance, CA, USA   **Area Size:** 106,000 sf

项目名称：萨奇广告公司 设计师：舒宾＋唐纳森建筑师事务所 项目地点：美国，加州，洛杉矶 面积：9847平方米

Focusing great attention on the main floor — centrally located on the third level — the architects designed a communal hub to bring staff together on one floor in a variety of configurations. This main space allows for impromptu collaborative moments.

From either entrance, the company's theme of "homing at work" is apparent. A spiral library brings to life the concept of a hearth and acts as a central gathering space. The walls form spirals in both plan and elevation toward its center. This central vortex of energy is further emphasised as the cone-shaped form from the ceiling beams down a soft glow of light toward this center space, known by the staff as "the pit".

Other communal spaces offered on this floor include a living room-like lounge area that mixes classic and retro furnishings by Vitra, large full-service kitchen and dining area, gaming room with pool table and plasma television, multi-media room with a ceiling-mounted projection screen, bar highlighted by a backlit orange wall, and a bamboo garden "backyard". This unusual space is defined by slender bamboo poles strapped together with steel bracing and a border of river rocks within a steel trough. Planters with lush greenery, fountain sounds, picnic table, and smaller tables with pillow-like seating, provide a tranquil oasis to inspire creativity. All levels offer private and group work stations, fostering communication and deterring competition. The senior staff have semi-private offices behind floor-to-ceiling glass walls.

设计师将重点放在第三层空间的打造上，在这里设置了一个集会枢纽中心，将所有的员工汇集到这一层。

无论从哪个入口进入公司，都可以体会到"办公如家"的理念。螺旋形状的图书室从壁炉设计中获得灵感，是员工交流集会的中心场所。

三层还设有其他公共空间，如休息室、厨房、餐厅、游戏室、放映室、多媒体室、酒吧以及竹子花园等。此外，公司内的所有楼层都设有单独的或是公用办公区，资深员工拥有半独立式工作室。

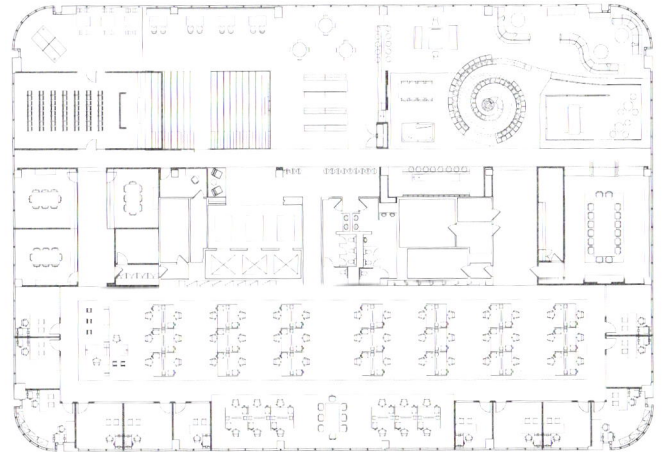

Office Lab

Lab办公室

Project Name: Office Lab　**Designer:** architects lab　**Location:** Brussels, Belgium
Photographer: Tim Vandevelde　**Time:** 2008　**Area:** 190 sqm

项目名称：Lab办公室　设计师：设计工作室　项目地点：比利时，布鲁塞尔　摄影：提姆·范德维德　完成时间：2008年
面积：190平方米

Architects lab located their new office at the ground floor of the block situated between the two courtyards. The large windows and fixed glass parts not only allow large amounts of light to brighten the office space but also make the transition between garden and workspace to be much less abrupt. Since most designs created by architects lab can be described as puristic, free of clutter, hence also the studio's name, they made sure also their office was an example of this design philosophy.

The flooring, 10 by 10 cemented laboratory tiles in black, is the main feature. They are used not only for the floor but also certain wall parts are covered in these tiles. The largest wall in the office features a large library wall in dark wengé wood, over 4 metres high. It houses all technical and other information needed in the office. These two main features are joined by simple but efficient office furniture, mostly also from the designer's hand. The best example is the light fixture which hangs in the office meeting/lunchroom, a simple but elegant construction in white tubular steel and TL lightbulbs. Overall the look of the interior is contemporary but it does also really refer to the atmosphere which must have been found in 1920's labs and offices.

设计工作室将新的办公室设在大楼的一楼，位于两个庭院之间。大面积的窗户和固定玻璃不仅将大量的光线引入室内，还让花园和办公区之间的过渡显得不那么突兀。工作室设计的大部分项目以简约、自由的混搭风格著称，如同工作室的名字。设计师们要将这间办公室打造成这一设计理念的典范。

地上铺着10x10的黑色水泥砖，就连部分墙面上也使用了这种砖。办公室中最大的一堵墙上，以黑色实木制成高达4米的档案柜，存放着设计工作需要的所有技术资料和其他信息。这里的办公家具大多出自设计师之手，虽然简单，却很实用。挂在会议室兼餐厅的吊灯就是最好的说明，它由白色的建筑用钢管和TL灯泡组成，简单却不失典雅。办公室的整体设计风格十分现代，但又借鉴了上世纪20年代办公室的设计元素。

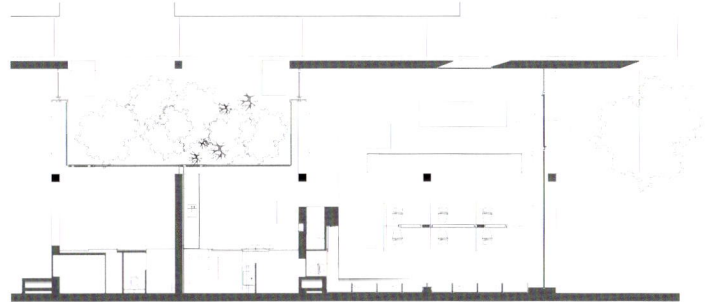

Turner Duckworth Offices

特纳·达克沃斯平面设计公司

Project Name: Turner Duckworth Offices   **Designer:** Jensen Architects   **Location:** San Francisco,USA   **Time:** 2005

项目名称：特纳·达克沃斯平面设计公司 设计师：延森建筑师事务所 项目地点：美国，旧金山 完成时间：2005年

Program & Requirements:

A graphic design firm purchased this concrete warehouse building in the Jackson Square District of San Francisco. The previous tenant had completely filled the building with a maze of closed rooms and private offices. The client's goal for the project was to convert the interior to an open office plan and uncover the double-height light-filled shell hidden by the previous build-out. The new interior was to reinforce the brand image and work style of the design firm and their own graphic identity.

Ideas & Design Objectives:

Red Glass Volumes

Taking red from the firm's graphic identity and mixing it with glass as a building material offering transparency, the architects proposed coloured translucent window film as the key ingredient in the spatial mix. Juxtaposed against the otherwise white minimal interior, these red room-volumes punctuate the space and provide a degree of privacy and separation between functions without subdividing the dramatic double-height volume of the building.

Cantilevered Glass Table and Glass Floor

A long cantilevered glass table is the main protagonist of the space. Centrally located in the open office area, it provides a multipurpose meeting space for the staff. The glass table hovers over a glass floor so that light from the large skylight directly overhead makes its way all the way down to the basement below. With a single gesture, a programmatic need, a light requirement, and vertical circulation (a staircase runs under the table) are all combined into a three-storey sectional composition.

Simple Construction, Soft Luminous Space

The existing building had a strong spatial and material character and where possible, it was left exposed. Only the rough concrete floor was covered with a white cementitious topping. The subtly off-white colour selected for paint, laminate furniture, and flooring, combined with the layered natural and artificial lighting, gives the space warmth unanticipated in its minimal composition.

设计要求

特纳·达克沃斯平面设计公司买下了位于旧金山杰克逊广场区的一个古旧仓库，并要求设计师将原有的封闭式小空间及单独办公区改造成开放式格局，并将天窗打开，以突出公司的品牌形象及工作方式。

设计目标与理念

红色玻璃空间

红色是公司的平面标识色，设计师将其与玻璃材质相结合作为主要建筑材质。红色玻璃空间在白色为基调的大环境中格外吸引眼球，在营造私密感的同时将不同的功能区分开。

悬臂式玻璃桌及玻璃地面

长长的悬臂式玻璃桌坐落在办公区正中央，构成了空间的主角，供员工会议使用。桌子悬浮在玻璃地面上，这样光线从天窗照射进来便可直接到达地下室。

简洁构造，柔亮空间

原有建筑无论是在空间布置或材质选择上都极具特色，这被大范围保留下来，仅有粗糙的水泥地面覆以白色漆面。此外，白色家具与地面同层次感分明的光线相结合，在简约的环境中带来了温馨气息。

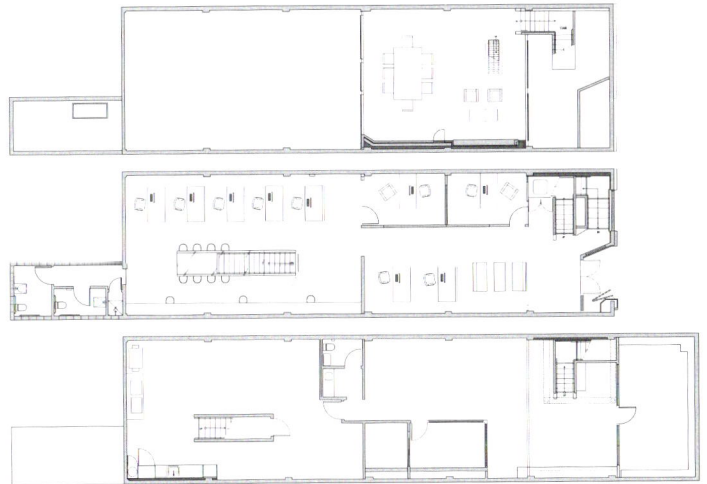

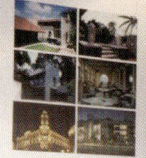

The Information Box

信息盒子

Project Name: The Information Box   **Designer:** neri & hu design and research office
Location: Shanghai China   **Photographer:** neri & hu design and research office   **Time:** 2009

项目名称：信息盒子 设计师：如恩设计研究室 项目地点：中国，上海 摄影：如恩设计研究室 完成时间：2009年

The information box was conceived as an apparatus to provide information for the apartments being sold in this prestigious development outside of Shanghai. The white oak clad wall of the box is also seen as a backdrop for the models and drawings displayed in the room. The wooden box has a series of doors, some of which slides out acting as divider for the space while other swings out with the inside showcasing pertinent information relevant to the different products being displayed outside. The conference rooms, offices, restrooms and back of house spaces are all located inside the box housing the team behind the project. The floors are clad in blue slate. Dark ceilings bring the calm atmosphere of space. White oak in the room and tender green in the corridor will stimulate people's vision and excite people. Regardless of all, the wooden box or the product itself is concerned, both played a very good background effects.

"盒子"主要功能即为提供位于上海开发区内一个新楼盘的在售信息。四周墙壁采用白色橡木覆面，为展示着的模型营造了完美背景。"盒子"四周安装有许多门，一些用作隔断，一些则用于展示相关产品信息。会议室、办公区、休息室及后勤室一应俱全。
地面采用蓝色石板铺设。天花饰以深色调，走廊是嫩嫩的绿色，让人们的感官得到充分享受的同时，也备感心情愉悦。除此之外，更为重要的一点是无论是对"盒子"本身，还是展品都起到了背景作用。

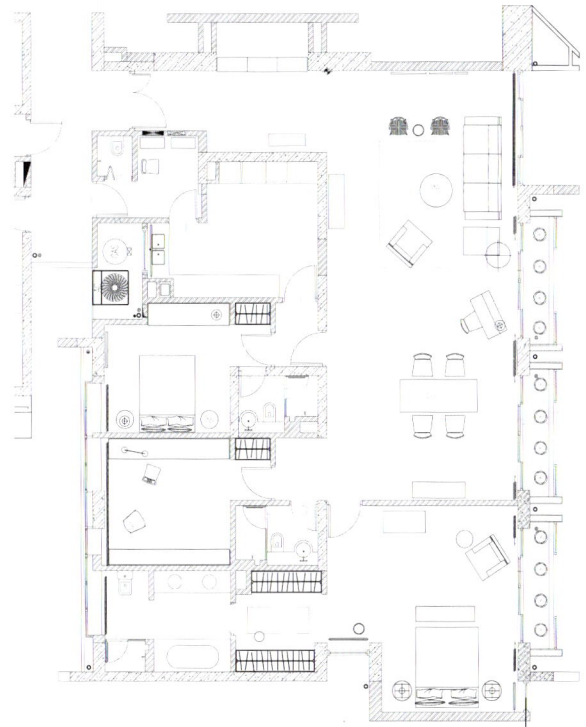

Lakeside Villa 湖畔佳苑

FOX Latin America Channel Base Offices

福克斯北美频道办公

Project Name: FOX Latin America Channel Base Offices **Designer:** Alberto Varas & Asociados, arquitectos Estudio Angélica Campi **Location:** Buenos Aires, Argentina **Photographer:** Sosa Pinilla **Time:** 2008

项目名称：福克斯北美频道办公 设计师：阿尔贝托·瓦拉斯联合设计公司 项目地点：阿根廷, 布宜诺斯艾利斯
摄影：索萨·皮尼利亚 完成时间：2008年

The new building's demands were: to obtain natural light for working areas and distribute the complex net supply of informatic, telephonic, and satelite services, the air conditioning, and the energy required by the new functions. Taking each one into working areas, offices and meeting rooms of the different layout areas.

So the problem became into an interior perforation game to get the entrance of the natural light and generate a spaciality according to the creative labours of each versatile sector, which means, a spacial system differentiated and interconnected between them, where inserted both unique faces of the building able to connect with the urban space. In the background facade, it was generated a vertical four-floor height garden, which delivers views and light to each floor of the building. And the front facade to the street, consists in a double facade, one operable and the other steady/fixed which regulates the entrance of light and reflects the building's technological character.

Passing from the front into the building's background it's a theoretic promenade which leads to the public world of technology, the sophisticated LED light and the data projections over the façade into the calmness of the garden and the silence of the internal working areas where people can briefly and illusoriness escape from the guessing and seductive digital machine in which insides works.

设计要求如下：确保办公区内光线充裕；均匀分配信息、空调及能源等元素，并使其融入到工作区、办公室及会议室中。

设计师从一个全新的角度诠释以上要求——将室内空间装饰转换成"打孔游戏"，为自然光线的入射建立"通道"；构思创意空间格局，突出灵活性。此外，他们在建筑背面打造了一个垂直花园，将光线及周围的景致带入到建筑内部。朝向大街一侧的立面由双层结构组成，调节光线的同时彰显了建筑特有的技术特色。

从正面走进建筑，一条人行道带领着走进技术的世界——精致的LED灯饰格外吸引眼球，继续前行便会体验到恬淡的花园氛围以及幽静的工作空间。

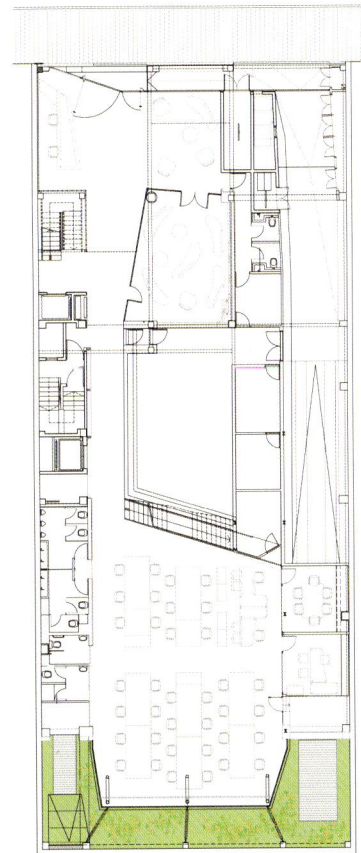

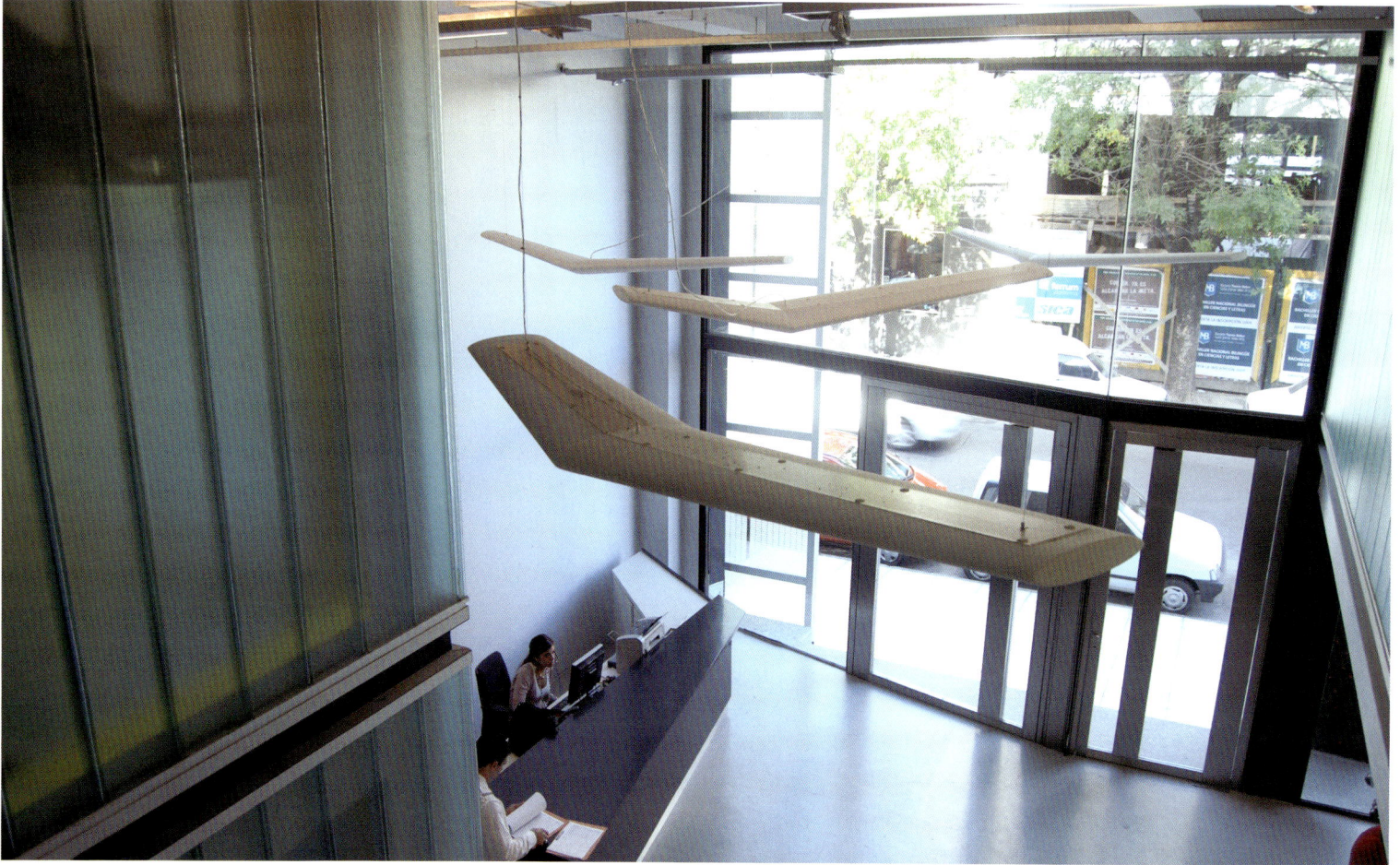

China Everbright Limited (CEL)

中国光大股份有限公司

Project Name: China Everbright Limited (CEL) **Designer:** Lam Cham Yuen & Lee Ming Yan, Olivia
Location: Shenzhen,China **Photographer:** Ulso Tsang Photographic **Time:** 2007

项目名称：中国光大股份有限公司 设计师：林湛元，李明焱，奥利维亚 项目地点：中国，深圳 摄影：奥索·普
完成时间：2007年

Being located on the 46th floor of a main building Admiralty, one would expect stunning window-side scenery, and the YO Design team cleverly incorporated these striking views into the office design. A lot of transparent glass has been used throughout the interior and this helps to share the uplifting views brought in from the expansive windows. In addition to the reception, there are five main rooms and these include one big conference room, two smaller conference rooms, a middle working space, and a guest meeting room. Off from the reception, the main guest meeting room is the most informal area, with a striped carpet and spotlight-like lighting fixtures adding an element of fun to the interior. Like the facing reception lounge area outside, this room is symmetrically arranged, with the furnishings mirroring each other on either side of the room. Nearby, the main working area features the same lined carpet as the guest meeting room, and this helps to stretch the space of the room. There are interesting selections of accompanying light fixtures to boot.

The main conference room seats 40 and although there are no windows, this room is mainly used for video presentations and slide shows.

项目位于金钟大厦主楼的第46层，每一个来这里的人都为窗外的美景而震撼，设计师们巧妙地将这些惊人的景色与室内设计融为一体。空间内部使用了大量的透明玻璃，像是将窗外的景观拉了进来。除接待室以外，公司共有五个主要的房间，包括一间大会议室，两间小会议室，一间中等大小的办公室，以及一间会客室。接待室外面的主会客室是公司中最随意的房间，条纹地毯和聚光灯为房间增添了些许轻松的气氛。和对面的接待休息区一样，会客室布置得十分对称，房间的一侧摆放着家具，另一侧是镜子装饰的墙面。附近的办公区使用了与会客室相同的条纹地毯，有助于拓展空间。有趣的是，你只要对着这里的灯踢一下，它就会亮起来。

大会议室设有40个席位。这里主要用于视频演示和播放幻灯片，因此没有窗户。

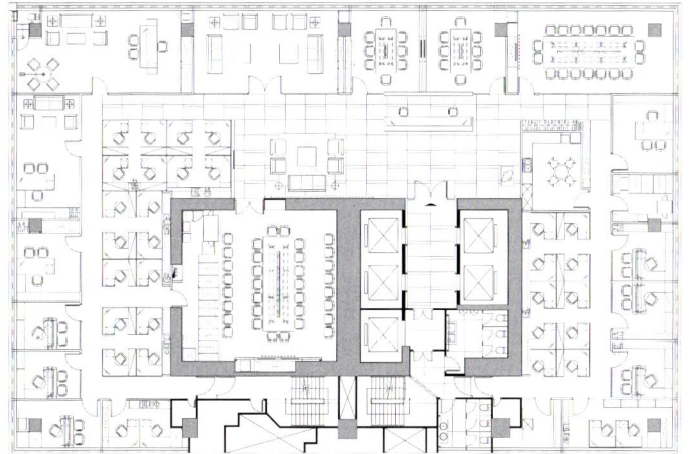

Ernst & Young University

安永大学

Project Name: Ernst & Young University   **Designer:** YO DESIGN LIMITED   **Location:** SShanghai China   **Photographer:** Ulso Tsang Photographi   **Time:** 2007

项目名称：安永大学 设计师：优设计有限公司 项目地点：中国，上海 摄影：曾宪成 完成时间：2007年

As business challenges become more complex, the need for international firm to call upon the widest spectrum of views and opinions to address them is crucial. Employed over 135,000 talented people in 140 countries worldwide, sharing the same value for work and commitment to quality. The open culture offers continuous personal and professional development. As a global organization, "we support individuals who wish to apply their skills and experiences in new environments… Because when people grow and succeed, it is the company that benefits."

Due to the reason, this global organization wishes to provide spacious and warm learning environment for their employees through the use of natural materials such as wood and travertine cream. The space appears as if it is almost nestled within its surroundings. The atmosphere created has a tendency to let people relax from the uptight and occupied life style.

Motorized pivot doors are featured in oak wood veneer creating a warm and natural focus in the lounge area, while full length grey and clear mirror wall finishing amplified the depth of the space. Designer accented with red furnishing, bold lighting fixture and bright colour soft cushioned operable wall paneling creates a dynamic, yet truly joyful learning experience.

对于国际化大公司来说，放宽视野开拓新领域至关重要，以应对不断复杂化的业务挑战。安永会计师事务所共有135000多名员工，来自140个国家，他们分享着共同的工作价值观，致力于为客户提供高品质服务。同时，极为开放的文化助于他们个人品德和专业水平的提高与发展。"作为全球化大公司，我们会为那些愿意加入到公司的人提供大力支持，他们的发展与成功会为公司带来更大利益。"

基于这一点，公司致力于为员工打造开阔、温馨的工作及学习环境。木材及石材等天然材质的运用达到了预期的效果，让员工在紧张的工作压力下得到放松。

大厅内，电动旋转门采用橡木板装饰，带来温馨自然的气息；灰色落地镜面墙则增添了空间的开阔感；红色的装饰元素、特色十足的灯饰共同营造了动感而愉悦的工作氛围。

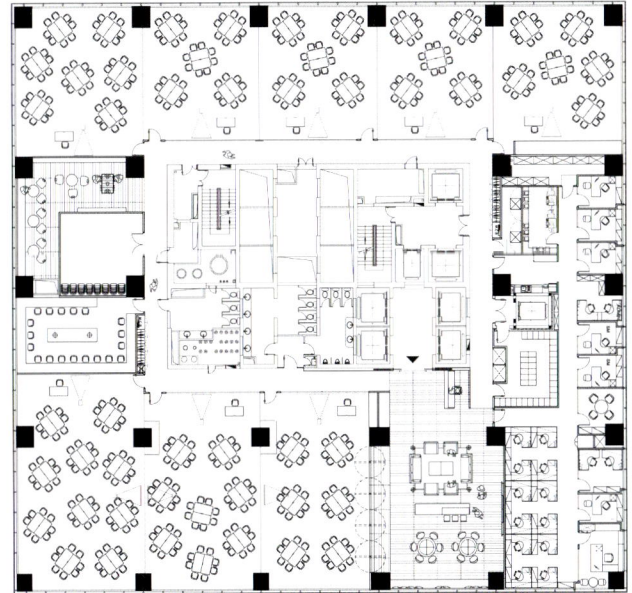

An International Firm— Shenzhen Branch

一家跨国公司的深圳分公司

Project Name: An International Firm—Shenzhen Branch **Designer:** Lam Cham Yuen & Lee Ming Yan, Olivia **Location:** Shenzhen **Photographer:** Ulso Tsang **Time:** 2006

项目名称：一家跨国公司的深圳分公司 设计师：林湛元、李明焱、奥利维亚 项目地点：深圳 摄影：奥索·曾
完成时间：2006年

Located near reception is a pair of meeting/conference rooms, partitioned from the rest of the office with floor-to-ceiling glass panels. Besides allowing penetration of natural light, the glazing also has a psychological effect, symbolising the client's 'transparent' and 'open' way of doing business. Dark cushioned carpets bearing a pattern of white irregular lines adorn the floors, while a few walls are paneled in chestnut-coloured leather to add a balancing impression of gravity. Custom-designed charcoal stainless steel door frames and skirting elements add to the rooms' sharp definition to the rooms. For the sake of flexibility, the two conference rooms can easily be combined into a single multifunctional space by simply opening up the wall panels that usually divide them.

A narrow, lengthy passage decorated with uplights, grew reflective mirrors and veneer-look, plastic laminated wall panels links the general office area with three training rooms, which can be merged into a volume large enough to accommodate up to 100 people. At the core of the building – and easily accessible from the general office – is a roomy "time-out" area where employees can eat and relax. To make its special status clear, the area is located on a raised wooden platform.

接待处旁边是两间会议室，由落地玻璃板把它们同其他办公室隔离开来。这样做除了可以让自然光照射进来，还起到心理暗示的作用，象征着客户"透明"和"开放"的经营理念。地上铺着黑色的地毯，上面布满不规则的白色线条，栗色皮革装饰的墙面为这里增添了些许严肃的气氛。定制的深灰色不锈钢门框和踢脚线鲜明地把一个个房间界定出来。两间会议室可以轻易地合并为一个多功能空间，只需打开中间的隔板即可，十分灵活。

一条狭长的通道连接着三间培训室和办公区。通道上方饰有顶灯、镜面和装饰板，两侧是塑料复合墙面。三间培训室可以合并成一个大房间，面积足以容纳100多人。办公区直通大厦的中心，这里是供员工吃饭休息的休闲区。这一区域采用加高的木制地板，以明确其特殊性。

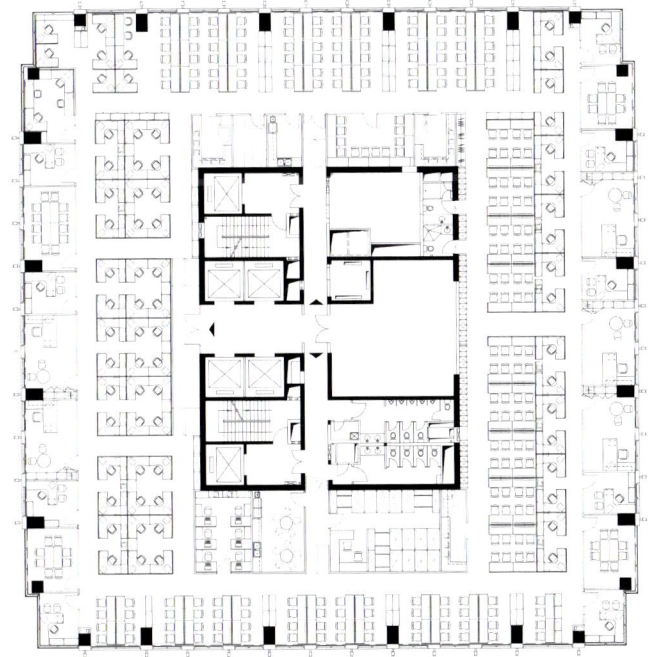

Chengdu Sales Office

成都销售处

Project Name: Chengdu Sales Office   **Design Firm:** One Plus Partnership Limited
Location: Chengdu, China   **Photographer:** Mr. Law Ling Kit   **Time:** 2007   **Materials:** Black tiles,
paint, stainless steel, gold colour mirror   **Area Size:** 467 sqm

项目名称：成都销售处 设计师：壹正企划有限公司 项目地点：中国，成都 摄影：罗灵杰,龙慧祺
完成时间：2007年 主要建材：黑地砖、油漆、不锈钢、金镜 面积：467平方米

All that can make an impact on the senses of a visitor as he crosses the threshold into a space is grist for the mill of instant assessment. Fairly or not, just as a cluttered and dusty living room will make guests think twice about a homeowner's standard of personal hygiene, a shabbily furnished office reception area will inevitably lead to doubts about a company's self-confidence and acumen.

Doubt of a rather more positive kind will likely entre the thoughts of visitors to this new office designed by One Plus Partnership, simply because it looks so little like an office. The spacious reception area glows with simple white surfaces and a subtle lighting scheme. A series of discontinuous curved partitions – some suspended in the air and some extending to the floor – divides the space into distinct-but-flowing zones, many of which are saturated with natural light courtesy of floor-to-ceiling windows. Details like artfully positioned black velvet sofas add a touch of softness to the scheme while preventing it from seeming too antiseptic.

Accordingly, the emphasis was on the architectural rather than the decorative, with Law and Lung endowing the space with a sharp-edged composition of simple forms executed in a palette of subtle-but-refined materials. An outstanding example of the designers' approach is seen in the discontinuous oval shape that circumscribes the office's central core. "The feature creates a smooth flow of presentation while at the same time adding to the impressive geometry of the area," Lung says. Its curve first emerges as part of the ceiling in the reception zone, drawing the visitor's eye deeper into the space where the feature transforms into a partition dividing a cashier's counter from the rest of the area. Finally, when it reaches the end of the volume, it rises to ceiling level to make way for a panoramic view through floor-to-ceiling windows.

不论公平与否,一个空间的实时印象,在甫进那一刻便已成形。一间凌乱不堪、满布灰尘的客厅,会令客人对主人的自信和能力产生质疑。

当然,初次印象也有肯定的。这间由壹正企划有限公司设计的办公室,便给访客意想不到的体验。宽敞的接待处,其纯白的墙壁和柔和的灯光令人目眩。弧形的隔板,或悬挂空中,或连接地面,把空间隐约地划分成多个区域。宽敞的落地玻璃窗,让和煦的自然光渗进室内。黑色的天鹅绒巧妙地置于空间中,为朴实的室内添上点点细节。

就此,两位设计师运用富于建筑性的风格,借由一系列精巧及优雅的物料,赋予空间线条分明的简单设计。其中最突出的是围绕中心部分的椭圆形结构。龙慧祺道："此结构令整个展示流程更顺畅,并为四周增添了几何线条。"椭圆形结构起始于接待处的天花板,进而把访客的视线带到更深的层次——直至收费处,此弧形结构跟地面连接成为该区的隔板。在空间的尽头,弧形结构再次恢复至天花板的高度,让访客享受落地玻璃窗外的景观。

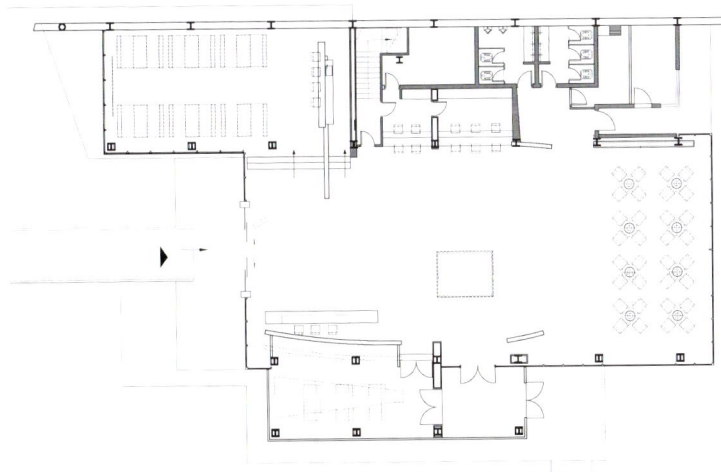

Machiya Office

町屋办公室

Project Name: Machiya Office   **Designer:**Hayakawa Hua Shi
Location: Niigata, Japan   **Photographer:** Koichi Torimura,Hirotaka Satoh   **Time:** 2007

项目名称：町屋办公室 设计师：早川华世 项目地点：日本，新潟 摄影：堂本光一 唐，佐藤润一 广
完成时间：2007年

The project "Machiya Office" was designed in Niigata (an old town in the north of Japan) in 2007. The total area of this project is 350.31 sqm (including 189.02 sqm floor area). The constructive structure of the office is steel frame; Machiya is a traditional townhouse along the street. The designers designed it as an office which is a base of widely spreaded business so it has the function of the face for that.

The shape of the site is narrow in frontage and long in depth. The designers proposed a two-storey auditorium with skylight that runs through the long axis of this tube-like building, which allows sunshine in and air out. That is called "walk through garden". Each room is divided up and down stair to face on "walk through garden". The auditorium and office are separated by a glass partition in a wooden frame and has a curved surface. The shape is formed by approximation for each room's area. To control the well, the glass is inclined. The bright and dynamic interior space creates an impressive corporate image for visitors, an important element of the project for the client. The facade references the black, stain painted wooden fences that make up the old town street.

项目坐落于日本北部一个古老的城镇——新潟，它建成于2007年，采用了钢架式建筑结构，其总建筑面积达350.31平方米（其中包含189.02的底层建筑面积）。
"町屋"是坐落于街边的一栋传统城镇居民住宅，之所以把它设计成一个办公室是因为：这里是商业广泛聚集的基地。
该建筑呈前端偏窄、整体狭长型。设计师们设计了一个双层的礼堂，令阳光铺洒整个建筑，增强空气流动，这被称之为"幽径花园"。每个房间都被划分成独立的个体，且楼下就面对着那个"幽径花园"。会议厅和办公室由一个镶嵌在木框里的玻璃隔板分开，形成了一个曲面——这种形状是根据每个房间的面积估算出来的，为了达到满意的效果，玻璃是倾斜着的。更重要的一点是：对于本建筑的所有者来说，其明亮且充满生机的内部空间使参观者对公司的形象十分深刻。其黑色的建筑正面以及被染了颜色的木质围栏，都为这条古老的街道增添了很多生机！

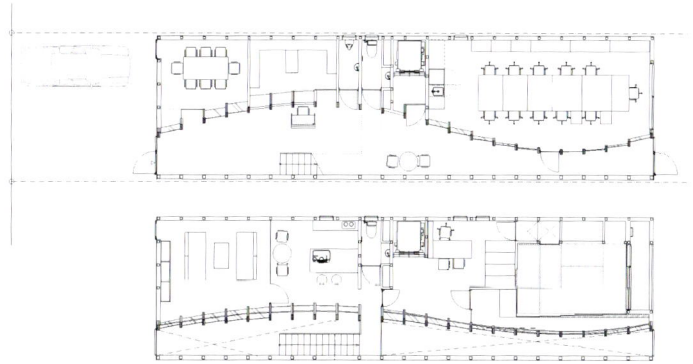

The Bekenstein Family
City Year Headquarters
for Idealism

CITY YEAR

City Year Headquarters for Idealism

"为理想奋斗"城市年鉴总部

Project Name: City Year Headquarters for Idealism   **Designer:** Elkus manfredi Architects
Location: Boston, Massachusetts   **Photographer:** Adrian Wilson   **Time:** 2007

项目名称："为理想奋斗"城市年鉴总部 设计师：艾尔卡司·曼弗雷迪建筑师事务所 项目地点：波士顿，马萨诸塞州
摄影：艾德里安·威尔逊 完成时间：2007年

City Year Headquarters for Idealism is the new permanent home base for the community service organization that inspired the creation of the federal AmeriCorps program. Elkus Manfredi Architects provided interior design services for this five-story, 55,000-square-foot signature space. The design tells the City Year story and supports the organization's brand, mission, and goals.

In 2004, Elkus Manfredi began working with City Year, real estate professionals, and the City of Boston to analyze building and development options. Once the appropriate building was identified, the team began to develop the concept for City Year's Headquarters for Idealism. To create a space with the same spirit and energy as City Year's young volunteers, a variety of graphics are incorporated into the design. Adorning everything from walls and glass partitions to floors and furnishings, inspirational messages and images fill the Headquarters with meaning and purpose.

The Lavine Civic Forum serves as a "theater of idealism," hosting corps-wide gatherings, guest speakers, and morning warm-ups. The room features large-scale video screens, moveable partitions and acoustic panels, whiteboards, a stage, and a pantry. Upper floors feature an open office plan that maximizes access to natural light. The second and the fifth floor are organized around collaborative spaces called "action pods".

The project is awaiting LEED-CI Silver certification.

"为理想奋斗"城市年鉴是一个为社区服务的组织机构，该组织为联邦美国义工团计划的设立提供了灵感。艾尔卡司·曼弗雷迪建筑师事务所为这个拥有5层楼、55,000平方英尺的建筑提供室内设计服务。设计师通过其设计将"城市年鉴"的故事娓娓道来，并很好地展现了该组织的品牌定位、工作任务和目标追求。

2004年，艾尔卡司·曼弗雷迪建筑师事务所开始与"城市年鉴"、房地产开发商和波士顿市政府合作，进行建筑设计和开发规划。确定了合适的建筑后，该团队开始通过设计彰显和发展"城市年鉴"组织为理想奋斗的理念。为了给"城市年鉴"组织中那些年轻志愿者创造一个与他们拥有同样精神风貌和青春活力的空间，设计中运用了一系列的图形元素，并使用具有特定意义的灵动文字和图片装饰了墙壁、玻璃、地面乃至家具。

"拉文公民论坛"作为一个"展示理想的剧场"，主要举办团体集会、特约报告会和晨会等。其空间的特点是拥有大屏幕投影设施、可移动的分区、音响控制台、白板、舞台和食品间。上层则设置了一间开放式办公室，以便最大限度地利用自然光线。2层和5层四周则分别设置了称为"行动集群"的团队合作空间。

该项目正接受商业建筑室内装修绿色评估体系银奖评估。

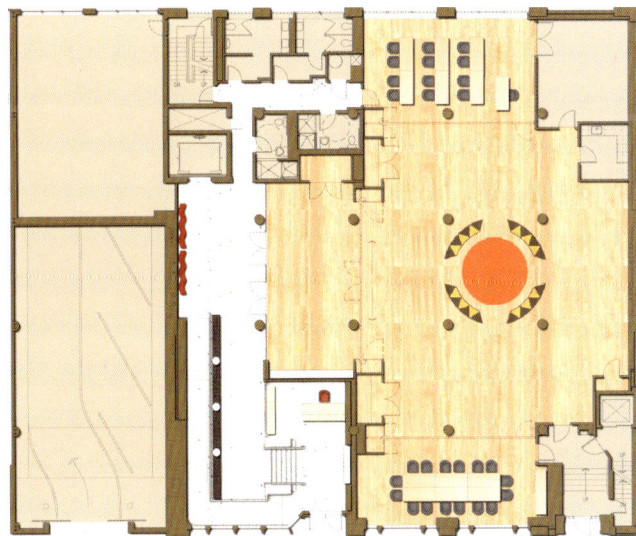

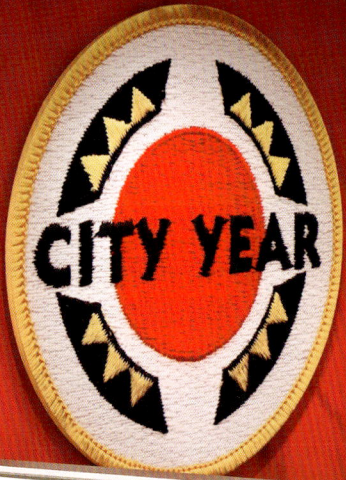

"You must **be the change** you wish to see in the world."

MAHATMA GANDHI

THE LAVINE CIVIC FORUM

imagine

PITW #52 Lead with ideas.

The first role of a leader is to lead with ideas, rather than rules or expectations. Allow people to get as excited as you are about those ideas. When people understand the "why" they will tend to quickly make the "how" happen.

"be the change you wish to see in the world."

MAHATMA GANDHI

people. My **human**

Çağdaş Holding Office

卡达斯公司

Project Name: Çağdaş Holding Office   **Designer:** Nagehan Acimuz   **Location:** İstanbul, Turkey
Time: : 2006   **Area Size:** 420 sqm

项目名称：卡达斯公司 设计师：纳格汉·阿西谬兹 项目地点：土耳其，伊斯坦布尔 完成时间：2006年 面积：420平方米

Çağdaş Holding is active in many fields like construction, real estate, concrete, etc. The priority design criteria in the office area created with gross concrete walls, coexistence of wood and glass, natural stone floor and permeable wooden divider is transparency. Differentiating room dividers were made transparent and for aesthetic considerations and privacy, permeable divider panels were added to these transparencies .

卡达斯公司业务涵盖建筑、房地产、原料生产等多个领域，其办公空间设计强调通透性，选用高光混凝土墙壁、木材、玻璃以及天然石材地面、木质隔断等元素。房间之间的隔断同样采用透明材质打造，但不失私密性。

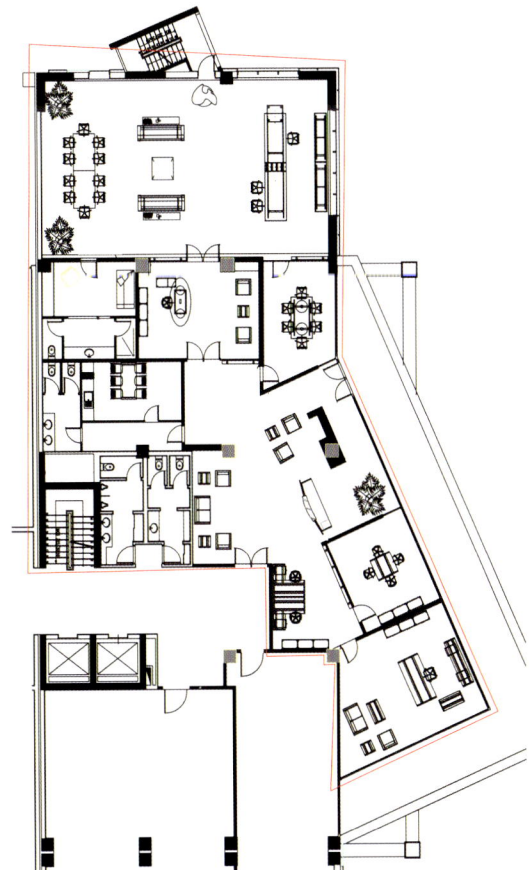

Meinl Bank

麦因尔银行

Project Name: Meinl Bank   **Designer:** ISA STEIN Studio   **Location:** Linz, Austria
Photographer: ISA STEIN   **Time:** 2007   **Area Size:** 2000 sqm

项目名称：麦因尔银行 设计师：ISA斯坦恩工作室 项目地点：奥地利，林茨 摄影：ISA斯坦恩工作室 完成时间：2007年
面积：2000平方米

To create a new business branch for a bank was a total new task for ISA STEIN Studio. Especially the bank is known for its exclusivity and tradition. What they tried to do was to bring these elements into a new design and they alao wanted that the innovation gets shown of course besides the traditional approach. Therefore they used a lot of detailed features like using their symbol of a blackamoor in surprising ways. Not only the client's zones are treated with elaborate materials, but also the recreation zones for the employees. Behind the big photo wall in the office space, you will find the kitchen and sitting area. So you have both sides for an approach. While working you are facing the image "buy low – sell high" and while being on a brake, you are seeing the image from behind.

The client zone is very flexible. The meeting rooms can be changed into one big meeting room, or left separate. The entrance area also has a nice place to sit into leather lounge chair, in order to create a more genuine atmosphere.

麦因尔银行因其独特的个性和传统而广为人知，因此为其设计全新的业务分支机构对于ISA斯坦恩工作室来说是一个不小的挑战。设计师们试图将所有的元素融合在一起，创新的手法同传统的手段相互融合。客户区和接待区同样采用精致的材料装饰；办公区内，宽大的图片墙壁后设置着厨房和座区。巧妙的设计为员工营造了不同的氛围——工作时，眼前呈现的是巨幅的图片，而休息时则可看到墙壁后面的风景。

此外，客户区设计强调灵活性——不同的小会议室可以合并成大的会议空间。入口处设有座椅，让人备感亲切。

DITTEL – Architekten

迪泰尔建筑工作室

Project Name: DITTEL – Architekten   **Designer:** d-arch   **Location:** Stuttgart, Germany
Time: 2008

项目名称：迪泰尔建筑工作室 设计师：d-arch设计工作室 项目地点：德国 斯图加特 完成时间：2008年

In a former horse stable in the Western part of Stuttgart DITTEL – Architekten found its new home. The new and modern bureau forms a contrast to the old building and offers a beautiful view over Stuttgart. The classical loft with an area of 230 square metres was completely renovated. The structure of the building was reconstructed and provided with a clear and clean alignment.

Different working areas (reception, modelling space, lounge, conference area, working tables) are separated by shelves which offer either an open or closed structure due to a flexible system. This system consists of a frame and varying cubes in three different sizes. In this way the look can be changed and the interior always meets the requirements of the daily life of the architects as it can be adjusted according to their needs.

The CI colours of DITTEL – Architekten, grey, white and light green, are reflected in the bureau. All pieces of furniture are finished in high-gloss, accentuated by a lime colour used in the inside of the cubes. Both the screed floor which is sanded, polished, and glazed/sealed and the 150 year old oak conference table emphasise the character of the building.

DITTEL – Architekten created a bureau where everybody likes to work, feels comfortable and is perfectly representative for its clients.

迪泰尔建筑工作室将新家落户在斯图加特东部地区的一个古老阁楼内，面积仅为230平方米，室内经过全面翻修。现代感十足的环境与古旧的建筑形成鲜明的对比，同时将斯图加特城的美景一览无余地收揽进来。

不同工作区（接待处、模型空间、大堂、会议室、工作台）通过橱架结构隔开，可根据需要随时调整空间规格，满足不同的需求。

公司的标识色——灰、白、绿在装饰中被引入进来。所有的家具全部经过高光处理，在橙色橱柜的衬托下，更加突出；地面经过磨砂、抛光、上釉等工序加工，同150岁高龄的会议桌共同彰示着建筑的特色。总之，公司为员工营造了一个格外舒适的环境，让他们乐此不疲地工作。

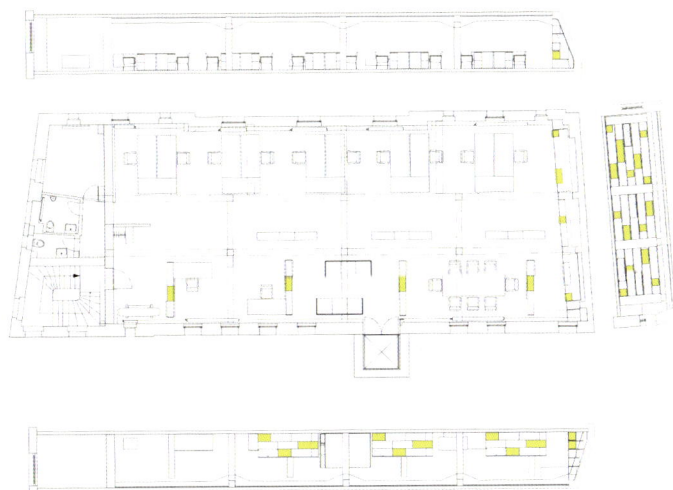

TIC

东京旅游信息中心

Project Name: TIC   **Designer:** Noriyuki Otsuka Design Office Inc   **Location:** Tokyo, Japan
Photographer: Noriyuki Otsuka   **Time:** 2009

项目名称：东京旅游信息中心 设计师：敬之大冢设计事务所 项目地点：日本，东京 摄影：敬之大冢 完成时间：2009年

Facing Nihonbashi Entrance Square of Tokyo Station, this centre is close to the shopping area crowded with long-established department stores and works as a hub for the access to the central Tokyo including Nihonbashi, Ginza, and Marunouchi business district as well as to the local areas throughout Japan conveniently. As the centre of tourist information in which visitors collect the new data with fun, it aims to enhance the internal/foreign tourism and to widen the tourist sites around Japan and the world starting from Tokyo.

The space was designed to make Tokyo's portal cool. The designer tried not to use traditional Japanese motif, like Zen space. The ceiling and the wall are painted in white; the flooring is in a special made gold plastic tiles. In red painted receptionist space, a counter made with white glass is placed with receptionists who speaks skillful in 4 languages (Japanese, English, Chinese and Korean), to allow them a detailed service. In the centre of the lobby, few laptop computer are placed on a round counter, visitors can access freely on TIC TOKYO Official Site with them. Under Japanese map, the self-service brochure rack is full of company and autonomy's brochure from all over the Japan.

The high ceiling who gives a kind of tension is decorated with photography of sponsors. Visitors seem to find contemporary designed TIC TOKYO Japanese. White color seems to mention Japanese castles, red color booth, Japanese flags. Well, Gold floor would mean `Chipangu (gold country)`told in `Description of the world` of the famous explorer, Marco Polo.

东京旅游信息中心朝向东京火车站桥本店入口广场而立，毗邻著名购物区，同时也是通往东京市中心的交通枢纽。作为为游客提供旅游信息的中心结构，这里旨在促进国内外旅游业的发展，同时开拓国内旅游新景点。

作为东京的门户，其实内空间的设计目标即为突出时尚新奇感。因此，设计师摒弃了传统日式主题（如禅屋风格）。天花和墙壁全部采用白色粉饰，地面则铺设着特殊制作的金色塑料瓷砖。接待区饰以亮丽的红色，白色的玻璃接待台前站立着热情周到的服务人员。大堂中心处，几台笔记本电脑摆放在圆形台面上，供游客使用。

此外，极高的天花给人张力十足的感觉，上面装饰着赞助人的画像。在这里，游客们会发现一个现代风格的日本，白色意指日式城堡，红色则象征着国旗，那么金色也许就是源自马可·波罗的《马可·波罗游记》吧！

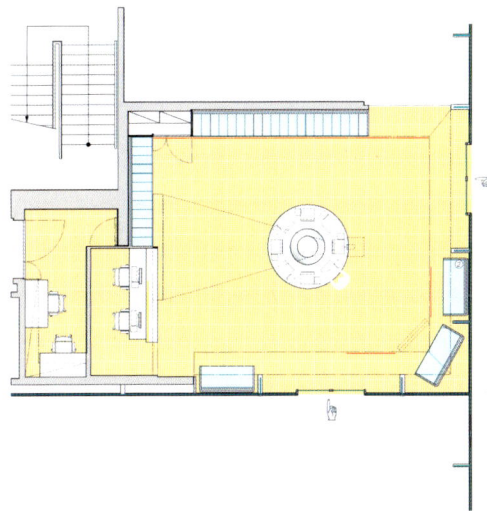

Shiyuan restaurant

食源餐厅

Project Name: Shiyuan restaurant   **Designer:** Lispace design studio,LiJia   **Location:** Beijing, China
Photographer: HanGao   **Time:** 2008

项目名称：食源餐厅 设计师：贾立/立和空间 项目地点：中国，北京 摄影：韩高 完成时间：2008年

Shiyuan Restaurant, located in the urban and rural trade centre of Beijing and famous for Sichuan cuisine, has a great number of regular customers. With the rapid development of its surrounding region, the owner decided to completely renovate the interior with the requirement of combining local elements of Sichuan. Finally, the designer proposed the theme of "Extraordinary Mountains and Rivers in Sichuan".

 With 520 square metres in total, the restaurant is composed of two parts: formal dinning room and fast food area. The designers broke the original closed layout with a full tempering glass wall in place of the concrete wall separating the restaurant and the department store beside. Moreover, part of the space is complete free of any kind of partitions to make it more spacious and transparent.

In the formal dinning room, the design shows great concerns in maintaining the privacy of the customers. From this point, the designers employed different dividing structures, for example, the aluminum sheets dangling from the ceiling. In addition, the floor and the ceiling are respectively covered with black Vitrified tile and white plaster in order to highlight other decorative elements. The multifunctional dinning area is specially designed for different events.

The fast food area, however, addresses fluidity, high speed and communication. The designers imaginatively play with "a piece of paper" (space) and fold it into different "parts". The counter, the desk and the ceiling are all covered with white artificial stones.

食源餐厅在北京城乡贸易中心的老店已经5年了，经营的川菜小吃一直吸引着很多固定的老客户，随着城乡贸易中心的升级换代，餐厅也决定进行一次彻底的改造，但在室内设计上要求设计师融入四川的元素。改造的面积为520平方米，分为正餐厅与快餐厅两大部分。经过设计师和业主的充分沟通后，确定了设计概念要表现"巴山蜀水"这一主题，而抽象与形象的思考便对设计的元素起到了决定性的作用。

空间整体的设计打破了原有封闭式的布局，餐厅和商场之间的隔墙换成了落地的钢化玻璃，局部采用了无隔断的虚拟空间，使得客人更直接的看到餐厅的内部，在不经意间走进餐饮的内部空间。餐厅更加通透，更易于交流，不再是一座封闭的孤岛。

正餐厅的区域划分充分考虑到了用餐客人的隐私性，座位的摆放，客人面对的场景都有所设计。不同的区域间采用了同样特殊的隔断方法：写意山水的悬吊铝片折条。这些装置艺术品在灯光的配合下营造了整个用餐空间波光淋漓的色彩，与主题紧密相扣。为了纯粹的表现主题，地面与天花没有做过多的装饰，地面采用了黑色玻化砖，天花为常见的纸面石膏板白色乳胶漆。多功能用餐区是为了方便城乡贸易中心内不同的公司举办活动要求而设置的，深色丝帘和铝片装饰隔断使简单的空间增添了一丝华贵。

与正餐厅相反，快餐厅的空间设计更体现了流动性、快节奏和人与人的交流性。从地面到桌面上升到天花后在折回地面，设计师是在玩笑着用一张"纸条"，折叠出了一个让人意想不到的自由伸展的空间。而这种折叠手法又和正餐厅里的铝片装饰同出一"折"。不同的是在空间中穿梭的吧台、桌面及天花采用的是白色人造石材料。墙面巧妙地用镂空板和绿色涂料表现溪中飞舞的水草，使整个空间在视觉上更具有层次感和冲击力。

餐厅改造的目的达到了：让客人游走于一个充满遐想与灵动的餐饮空间，享受着美食与设计带来的快乐。

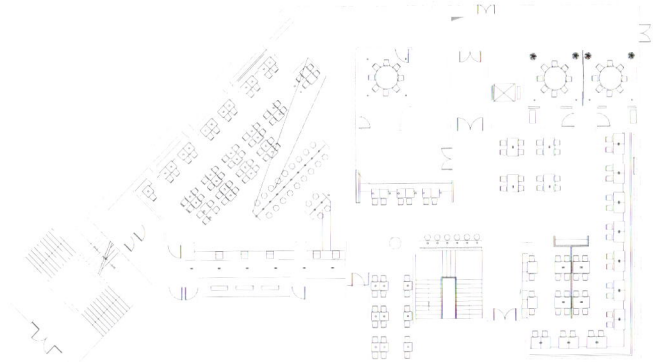

Maedaya Bar

马依达雅酒吧

Project Name: Maedaya Bar   **Designer:** Architects EAT   **Location:** Victoria, Australia
Photographer: Derek Swalwell   **Time:** 2008   **AreaSize:** 120 sqm

项目名称：马依达雅酒吧 设计师：EAT设计事务所 项目地点：澳大利亚，维多利亚 摄影：德里克·斯沃韦尔
完成时间：2008年 面积：120平方米

This project demonstrates the possibility of using ordinary recyclable material for hospitality projects without compromising the sophistication of the food and service.

Traditionally sake is bottled in wooden casks and secured with ropes. Current commercial method of bottling sake is similar to that of red or white wine. Designers' interest in sake bottling lies in the bounding of the cask using ropes. Thereafter they chose to investigate and translate their interpretation of 'bounding' with the use of Manila ropes. The ropes held in tension at specific points form a shape of a house or a hut. A house, whether it be a tea house or sake house is a sacred place in traditional Japanese times. It is a place where people drank in harmony.

The first floor function room is a space for self service grill. The essence of the space is in starck contrast with the ground floor. Here designers wanted to achieve a sense of humbleness: white washed walls, Japanese black stained timber flooring, simple timber benches and raw stainless steel canopies. There is no attempt to apply any embellishments because designers wanted the cooking to embody the purpose of this space.

这个项目表明，利用普通的可回收材料同样能创造出完美的效果，并不会降低档次。

传统的储酒方式是将酒装在木桶里，外面用绳子加固。现在红白葡萄酒的储藏方式与之类似。设计师们对此很感兴趣。他们对传统进行创造性的解读，用马尼拉绳索来诠释"捆绑"的概念。绳子被拉直后，沿着房子内部的形状分段固定。用这样的房间作传统日式风格的茶馆或清酒屋，都会给人以庄重的感觉。人们在这样的场所畅饮，心情也会愉悦起来。

二楼是自助烧烤区，风格与一楼截然相反。设计师将这里设计成平易近人的感觉，采用白色的墙壁、日式黑色木地板、简单的木板长凳和不锈钢排烟罩。这里没有多余的装饰，烹饪才是最重要的。

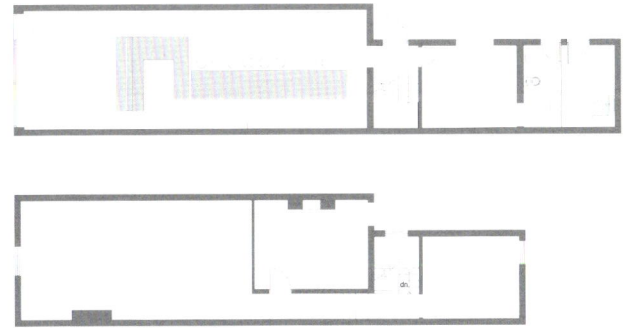

Adour Alain Ducasse

阿杜尔·艾伦·杜卡斯餐厅

Project Name: Adour Alain Ducasse   **Designer:** Rockwell group
Location: California US   **Photographer:** Bruce Buck ,Eric Laignel   **Time:** 2008

项目名称：阿杜尔·艾伦·杜卡斯餐厅 设计师：罗克韦尔设计组 项目地点：美国.加利福尼亚 摄影：布鲁斯·巴克，
埃里克·莱涅尔 完成时间：2008年

The main dining area is surrounded by wine displays in temperature-controlled armoires. Given the existing turn-of-the-century architecture of the St. Regis, special care was taken to create an environment that is both contemporary and dynamic, and sensitive to the building's architectural and historic context.

Elegant vitrine portals have been inserted into the existing stone wall entry and corridor to open the space up to Astor Court. In the main dining room, the existing architecture is covered in silver leaf and the walls are sheathed with a backlit, seeded glass veil etched with an abstract grapevine pattern, allowing one to look through the new to the old. Leather-covered furnishings in wine tones help create a warm, glowing atmosphere.

The main dining area has three private dining rooms. Two oval-shaped rooms with mirrored ceilings and fabric-covered walls are located in the corners of the back of the dining room. The oval rooms will be visually connected by a five-panel mural by the acclaimed New York artist Nancy Lorenz.

A third rectangular dining room features 50 leather and bronze wine lockers that enable guests to store either 12 regular bottles or 6 magnums of wine, as well as personal wine accessories, and a custom-designed album to record and preserve wine labels.

餐厅的四周是用于展示葡萄酒的温控酒柜。鉴于瑞吉酒店的悠久历史，设计师为餐厅精心营造了现代动感的氛围，又根据酒店的建筑及其历史背景加以调整。

设计师拆除了原有的石墙，装上优雅的橱窗，从餐厅里能看到对面的阿斯特阁咖啡厅。餐厅内贴着银箔，墙壁上打着背光灯，玻璃上刻着抽象的葡萄图案，提醒着顾客这里过去的样子。暗红色的皮质家具营造出温暖热情的氛围。

餐厅内有三个包间。其中两个包间为椭圆形，位于餐厅后面的角落里，包间内装有镜面天花板和布艺墙面。纽约著名艺术家南希·劳伦斯创作的五幅连环壁画将这两个椭圆形的包间连接起来。

第三个包间为长方形，内有50个皮革和铜制的葡萄酒柜，顾客可以在这里存放12瓶正常容量或6瓶大容量的葡萄酒，以及自己的葡萄酒配件，这里还有特别定制的画册，用于记录和收藏葡萄酒的标签。

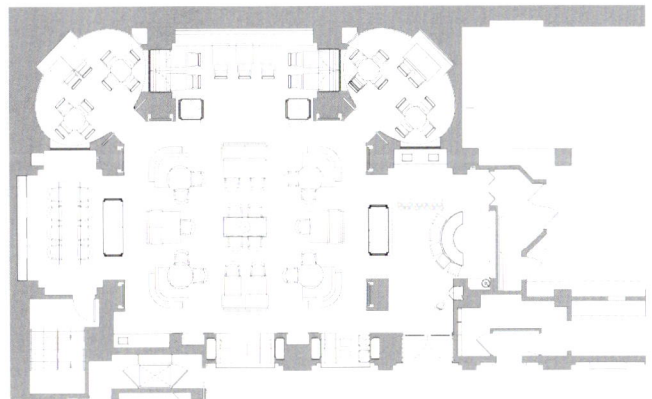

Adour at The St. Regis Washington, D.C.

华盛顿圣瑞吉酒店的阿杜尔餐厅

Project Name: Adour at The St. Regis Washington, D.C.   **Designer:** rockwell group
Location: Washington   **Photographer:** Bruce Buck   **Time:** 2008

项目名称：华盛顿圣瑞吉酒店的阿杜尔餐厅 设计师：罗克韦尔设计组 项目地点：美国，华盛顿 摄影：布鲁斯·巴克
完成时间：2008年

The designer preserved the original wood paneling and the base of the bar while updating the 84 square metres space with modern touches of stone and matte black lacquer, subtle metallic wall finishes, and furniture upholstered in sumptuous violet hues. The bar area is strewn with violet, grey, black and cream materials and finishes, in contrast to the 153 square metres main dining room, which is dominated by shades of white, black, and silver. The main dining room is designed with the same sensitivity to its historic heritage. Designer preserved the beautiful ceiling-height arched windows lining one wall letting in an abundance of natural light, the original woodwork throughout the space, and the existing 16 foot-high beamed ceiling, which also continues into The Bar. The landmark ceiling is composed of carved wooden flowers and decorations, providing an elegant classical contrast to the modern design elements that designer added, such as the 20-foot long Chesterfield-style dining banquettes along one wall upholstered in opalescent white leather and illuminated by rock crystal lamps. Other new additions to the space include three curved wood-wrapped dining niches that contain sleek black, channel-tufted banquettes and metallic wall coverings. Wenge and imbuye wood architectural details are juxtaposed against the matte black lacquer added to the existing woodwork in the space.

设计师保留了酒吧原有的木质镶板和其他原本设施，在84平方米的空间内，利用石材、黑色哑光墙面、精美的金属墙饰和豪华的紫色家具，充分提升了空间的现代感。酒吧所使用的材料色彩丰富，有紫色、灰色、黑色和淡黄色，紫色的家具，与153平方米主要采用白、黑、银三色的主餐厅形成了鲜明对比。
主餐厅的设计也是在其悠久的历史背景之下展开的。设计师保留了漂亮的拱形落地窗，以获得充足的自然光，还保留了原有的木制品和一直延伸到酒吧的16英尺高的天花板。天花板上有雕刻而成的花卉图案和精美饰品，有着优雅的古典美，与后期加入的现代元素形成对比。一张20英尺长的英式坐席依墙而设，外表采用乳白色的皮革，上方以水晶灯照明。设计师还用弯曲的木板圈出三个包间。包间内采用光滑的黑色植绒坐席和金属色壁纸。在原有的木制品中加入这些木制装饰，与空间内的黑色哑光墙面相映成趣。

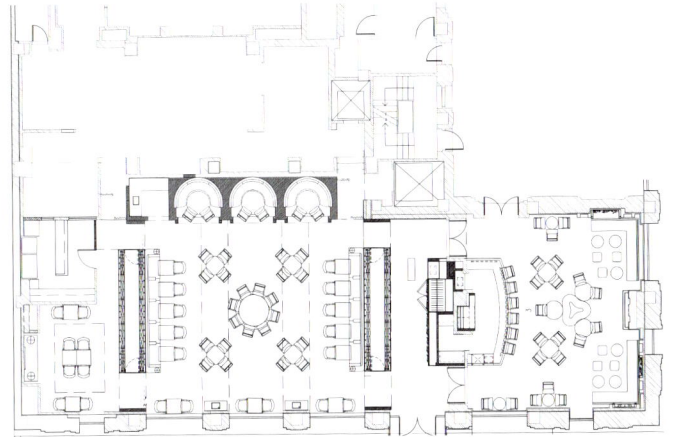

DOS CAMINOS

Matsuhisa

马苏希萨餐厅

Project Name: Matsuhisa   **Designer:** rockwell group   **Location:** Mykonos. Greece
Photographer: Vagelis Paterakis   **Time:** 2007

项目名称：马苏希萨餐厅 设计师：罗克韦尔设计组 项目地点：希腊，米克诺斯 摄影：瓦格列斯·帕特拉基斯
完成时间：2007年

The designers converted the restaurant which was only a sushi counter into a full restaurant, incorporating the clients' specialised menu and iconic accents in the interior, furniture, and décor. The space includes a main interior space, one garden area with five bougainvillea vines growing over it, and two terraces, all used for dining.

To reach the main dining room and terraces, guests enter through the garden, which includes a new white marble sushi counter. Throughout each area, the designers have renovated the seating. They also designed hand crafted matchbook screens that Silverhill fabricated which wrap the entire interior space and create the ceiling and walls. Silverhill also painted images of cherry blossom trees on panels below the screens. In front of the matchstick screens are light fixtures made by Bocci (Italy), which are cast glass balls with halogen lights housed in cylindrical voids in the middle.

The back terrace is a classic lounge/bar with low seating, small scorched ash tables, and another marble bar. The second terrace is another outdoor dining area. Both terraces have beautiful vistas looking downhill over the town of Mykonos.

设计师将这家原本只做寿司的小店转变成一家菜色齐全的餐厅，并将这里特别推荐的菜系和特色食物融入室内设计之中。餐厅包括一个大房间，一个花园，五株九重葛的藤蔓生长的十分茂盛，还有两个露台，全部作为用餐区。

顾客要到达主餐厅和露台，必须先经过花园，这里增设了一个白色大理石的寿司柜台。设计师对每一个领域的座椅都进行了重新设计。他们还设计了火柴盒造型的屏幕，采用手工制造并由"银山"品牌组装而成。这些屏幕充斥着整个室内空间，包括天花板和墙壁。屏幕下方的樱桃树也出自"银山"之手。屏幕前的灯具来自意大利的"波奇"品牌，这些水晶球形状的卤素灯安置在餐厅中央的圆柱形空间中。

后面的露台是一个古典风格的休息室和酒吧，这里有低矮的座椅、深灰色的小桌子和与花园相同的大理石柜台。另一个露台也是户外用餐区。从这两个露台都能欣常到米克诺斯城的美丽景色。

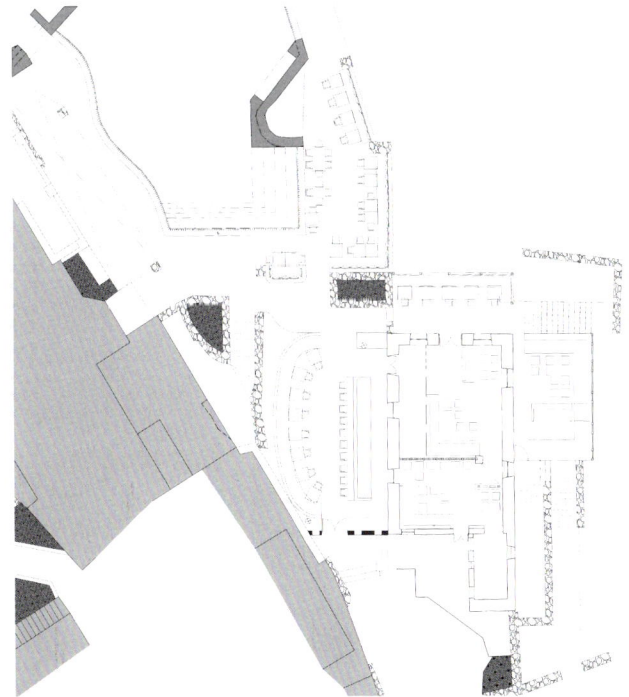

-ALL·DAY-
~ STARTERS ~

Kicked·Up Caesar
Honey Cornbread Croutons, creamy Chipotle Dressing

Jailhouse Red Chili
Smoked Brisket, Pinto Beans

Chicken Wings
-Rubbed & Fried-
OR
-Slathered-
with
·Chipotle·Raspberry·

Frank's Redhot

Bottle Caps
Beer—Battered fried Jalapeño Slices, Ranch Dressing

The Hutto Texas Wedge
Iceberg, candied Pecans, Bacon, Blue cheese Dressing

Wildwood Barbeque

怀尔德伍德烧烤屋

Project Name: Wildwood Barbeque **Designer:** rockwell group **Location:** Sao Paulo, Brazil
Photographer: Eric Laignel **Time:** 2008

项目名称：怀尔德伍德烧烤屋 设计师：罗克韦尔设计组 项目地点：巴西，圣保罗 摄影：埃里克·莱涅尔 完成时间：2008年

Rockwell group show us a new BBQ by their new concept, which leads us to a new kind of BBQ in the city. The inspiration opens a new door for traditional concept. Wildwood BBQ is a 325 square metres space which is an intersection of both the rustic and the industrial. While the rustic theme is underscored by the restaurant's expansive wood-timber ceiling, the industrial aesthetic is emphasised by the use of steel-frame garage doors filled with nicotine-stained glass panes above the 50-foot bar and the original concrete floor. Blackened steel details, found on much of the furniture and architectural elements, serve to further illustrate the industrial theme.

 At Wildwood BBQ, designers embraced the challenge of re-purposing ordinary and found materials in creative ways. The restaurant also implements several raw and environmentally-conscious materials, such as plywood-clad walls, reclaimed wood, and even a bar top made of 100% recycled paper. What results is the birth of a new kind of BBQ environment; one that is influenced by it's urban setting. This is how we do barbeque in the city.

罗克韦尔设计组以全新的设计理念向人们呈现了一个新型的烧烤屋，为人们展示了城市烧烤屋的新风范。灵感为传统的思路开启了一扇新的大门。怀尔德伍德烧烤屋占地325平方米，是乡村风情与工业感的混合体。餐厅的木质天花板突出了乡村的感觉，而钢架车库门，50英尺长的柜台上镶着的咖啡色玻璃和水泥地面，则强调着工业的美感。很多家具和设计元素中都以黑钢作装饰，进一步强化了工业主题。

在怀尔德伍德烧烤屋，设计师面临着一项挑战，就是发掘普通材料的新用途，并以创造性的方式应用在餐厅之中。餐厅还采用了许多原生态的环保材料，如胶合板铺就的墙壁，可再生的木材，柜台面甚至全部用再生纸做成。于是一个新式烧烤屋诞生了，它将对城市环境产生影响。我们在城市里就是这样烧烤的。

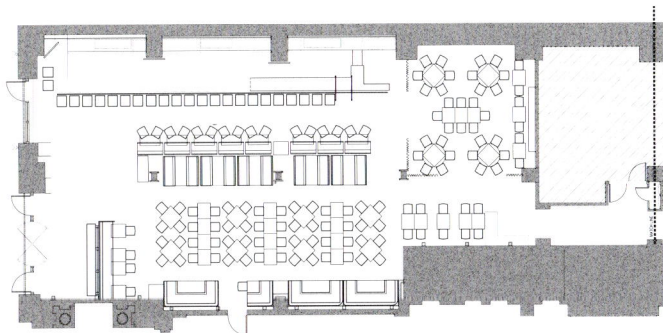

Tiandi Yijia – Restaurant

天地一家餐厅

Project Name: Tiandi Yijia – Restaurant　**Designer:** Mauro Lipparini　**Location:** Shanghai China
Photographer: Gionata Xerra　**Time:** 2007

项目名称：天地一家餐厅　设计师：Mauro Lipparini设计事务所　项目地点：中国，上海　摄影：吉奥纳塔·谢拉
完成时间：2007年

Light and shadow define the mood of a space while silently interpenetrating one another in order to engage the occupant of the space. The magical key to the Tiandi Bund n° 6 project is "luxury". The use of contemporary light fixtures and indirect light sources increase the interplay of light and shadow: light is projected though cuts and slits in the false ceiling patterns, and filtered through crystal and glass.

The "set design" of the restaurant is developed as follows: on the left-hand side you will find a large wall clad with natural travertine stone; on the right-hand side you will find a lounge-type area that has the warm, comfortable feeling of a living room outfitted with sofas. In the central area behind the staircase, you can catch a glimpse of the clear, luminous foyer articulated by crystalline lighting all of which create a jewel-like feminine environment. The foyer has a gazebo-like design with a vaulted ceiling made up of light-green glass pilasters and glass panels on three sides. The fourth side is composed of a sculpted travertine stone wall. From the foyer, one enters the Grand Hall, divided fundamentally into three naves: the central area expresses a magical and mysterious use of light. Both sides of the Grand Hall contain two long "niches" composed of pleated fabric and indirect lighting placed above the false ceiling, thus creating a whole new "adventure" of light — curling shadows, softly lit rippled dunes in the desert.

天地一家餐厅位于外滩6号，主要设计理念即为"彰显奢华"。光影是打造空间氛围的基础元素，相互作用吸引着顾客们陶醉其中。基于这一点，设计师充分运用现代照明设备，借以打造独特的光影效果——光线照射在天花板的凹凸图案上，经过水晶和玻璃过滤，然后洒落下来。

餐厅的空间布局简洁清晰：左侧是一面由天然石灰石饰面的墙壁，右侧则是休息区，里面放置着沙发，洋溢着温暖舒适的气息。楼梯后侧的中央区域是明亮的大厅，晶莹闪烁的光线营造了奢华柔和的氛围。拱形的屋顶由浅绿色玻璃板材打造，如同凉亭一般。穿过大厅便可达到就餐区，分成三个部分：中心区强调魔幻意蕴；两侧则是长形的空间，在灯光运用下给人以全新感受。

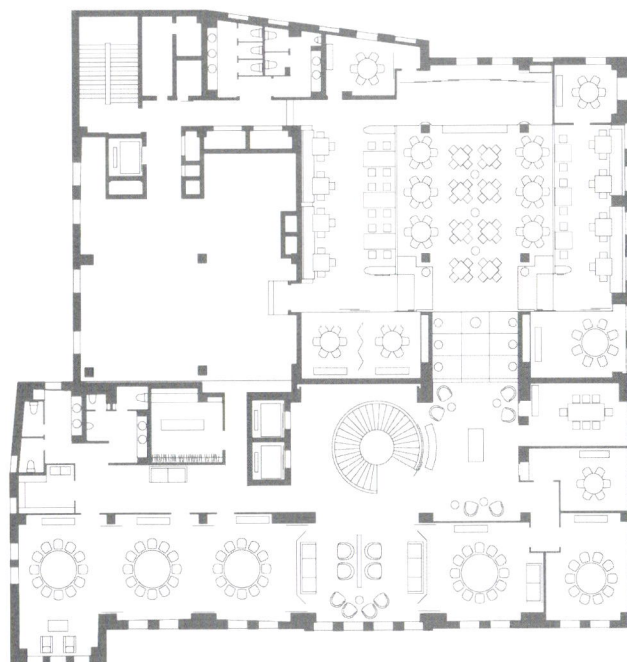

BEI — Asian Restaurant

"北" 一 亚式餐厅

Project Name: BEI — Asian Restaurant   **Designer:** Lyndon Neri and Rossana Hu
Location: Beijing, China   **Photographer:** Derryck Menere and Michael Webber   **Time:** 2008

项目名称："北" 一 亚式餐厅 设计师：林登·内里 胡如珊 项目地点：中国，北京 摄影：德里克·梅奈莱 迈克尔·韦伯
完成时间：2008年

Walking into Bei, one is surrounded by a forest of trees – wooden members physically and visually screening those within from the outside. Passing through the second screen one enters a clearing, the Public Dining area above which bird lights dance, seemingly caged within the screen. Facing the Raw Bar a large mirror behind the chefs allows the patrons to celebrate the art of the cuisine.

Visible beyond the clearing are five white boxes, carved within to house the private dining rooms. Windows above lead one's eye to urban scenes, and small openings in reflective compositions on the wall allow the diners to view into the other's space.

走进餐厅，仿若进入森林一般被包围在树丛中。穿过第二道屏障，便会到达一片空地——公共就餐区，头顶灯光闪烁，好似在笼子中跳舞的小鸟一样欢快。吧台对面是一面大镜子，可以看到厨师们忙碌的场景，品味烹饪的艺术。

公共就餐区旁边是5个白色盒子结构——单独就餐区。目光穿过墙壁上的窗户和小孔，便可看到其他空间的景象。

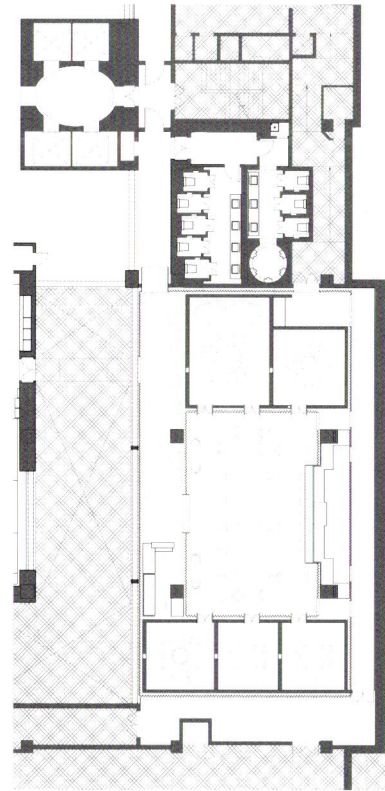

Sureno (Mediterranean Restaurant)

Sureno — 地中海餐厅

Project Name: Sureno (Mediterranean Restaurant) **Designer:** Lyndon Neri and Rossana Hu
Location: Beijing,China **Photographer:** Derryck Menere and Michael Webber **Time:** 2008

项目名称：Sureno — 地中海餐厅 设计师：林登·内里 胡如珊 项目地点：中国，北京 摄影：德里克·梅奈莱 迈克尔·韦伯
完成时间：2008年

Concrete walls define the more public, circulatory areas of Sureño while the spaces beyond with their surfaces of warm woods and blue olive grove patterns remind one of the Mediterranean. Winding through the entry, the hearth of the restaurant is revealed – the Pizza Oven – with its surrounding counter, allowing patrons to gather and watch the chefs at work. Breaks in the ceiling allow light to stream through and the ceiling's slope upward in the public dining area emphasises the view of the Sunken Garden. The wooden floor extends into the Sunken Garden to become a deck and the water pools within the restaurant and outside at the Sunken Garden ensure their connection to each other and the Sea is not missed.

餐厅公共区内墙壁采用水泥材质饰面，与其相邻的区域则以暖色木材及蓝色橄榄枝图案装饰，让人不禁联想到地中海风格。走过入口，便可看到餐厅的"主角"—比萨烤箱，四周摆放着吧台，客人们坐在这里可以尽情欣赏大厨的烹饪表演。天花板上的开口便于光线入射，公共就餐区上方的屋顶呈倾斜状，突显了低洼花园（Sunken Garden）的景致。木质地板一直延伸到花园内，形成了室外露台，餐厅内以及花园处的泳池相互连通。

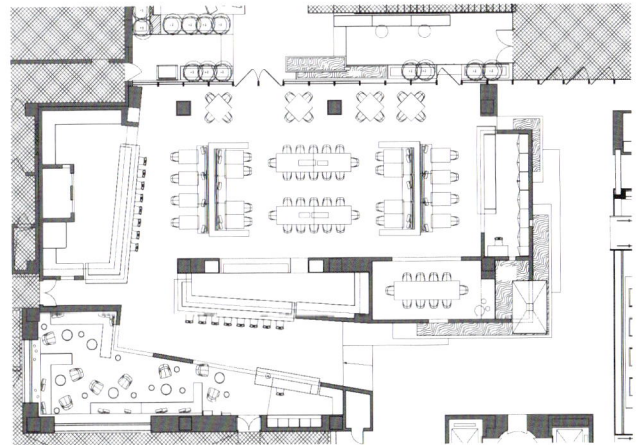

Cityscape Restaurant

城市景观餐厅

Project Name: Cityscape Restaurant **Designer:** Matt Gibson Architecture+Design Studio
Location: Melbourne, Austrilia **Photographer:** John Wheatley **Time:** 2008

项目名称：城市景观餐厅 设计师：马特·吉布森建筑设计工作室 项目地点：澳大利亚，墨尔本 摄影：约翰·惠特莉
完成时间：2008年

The design intent was to maximise the configuration of the existing restaurant space to provide for an effective educational layout whilst also making best use of the existing facilities which included views to, from and within the restaurant.

The existing restaurant (unrenovated since the early 80's) was located behind closed doors and had been accessed via a long dark corridor leading adjacently into the existing space. Once inside, the existing restaurant did however contain stunning panoramic views of the City of Melbourne skyline. These contextual elements informed the decisions to first, open up the restaurant space to the rest of the building internally via the use of a large glazed entry and display area , and second, to reposition the existing bar so that it orientated itself and the entire restaurant configuration toward the view of the city. Devices such as an angled mirror above the bar-top is used for student lessons to gain a 'birdseye' view of the bar top action, but it also reflects the view of the city from the bar meaning the city is in view from all directions.

Existing rooms within the restaurant are retained where possible and simply re-cycled for other functions. The existing operable wall spanning across the restaurant was retained and re-clad, and another operable wall was added to enable the flexibility to break up the space into 3 separate isolated workshop areas. With this wall closed the Private Dining & Meeting room is used for small conferences with a pull down projection screen, or it can provide an elegant makeshift 'Boardroom' for administrative or lunch meetings with external business entities. Alternatively, the restaurant space can be extended by 12 covers when the operable wall is open.

设计的目标是将原餐厅的空间进行最大化的重构，来提供既能满足高效学习的布局，又能充分利用原餐厅设备，包括餐厅内外的视觉体验。

原餐厅从80年代就没再翻修过了，坐落在一排关闭的大门后面，人们通过一条长长的黑暗走廊来到这里。但是原餐厅的视觉体验却非同寻常：一走进餐厅，整个墨尔本城市的全景尽收眼底。这些背景元素影响了设计策略：首先，通过镶玻璃的宽敞入口和展示区，把室内空间进一步向建筑的其他部分打开；其次，改变吧台的位置，使之引导人们的视线望向整个餐厅以至向外望向城市景色。吧台上方呈一定角度设置的镜子，让学生能"鸟瞰"吧台上的动作，同时镜子也折射出城市的景色，也就是说从各个方向，城市都尽收眼底。

餐厅里原来的房间尽量保留，只是简单地重新布置，改换了其他功能。原来跨越整个餐厅的活动墙也得以保留，墙壁经过重新覆盖，此外又增加了一面活动墙，来增加餐厅的灵活性，能把空间分成三个单独的工作间。用上这面墙，私人用餐或会议室就产生了，放下投影屏幕，可以开小会；同样也可以用作临时的大会议室，可以是管理层的会议，也可以是与其他外部的商务团体聚餐。如果把活动墙打开，餐厅空间可容纳12个桌位。

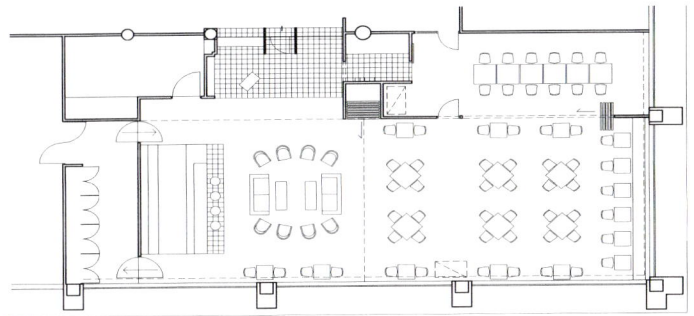

CITYSCAP

NEVY

聂维餐厅

Project Name: NEVY   **Designer:** Concrete Architectural Associates
Location: Amsterdam the Netherlands   **Photographer:** Ewout Huibers   **Time:** 2008

项目名称：聂维餐厅 设计师：混凝土设计事务所 项目地点：荷兰，阿姆斯特丹 摄影：艾沃特·胡贝尔斯
完成时间：2008年

Concrete was asked to redesign the formerly known restaurant Onassis. The restaurant is a new member in the succession of two existing restaurants; delicacy restaurant "envy" and wine bar "vyne". The facility for this location is a fish restaurant called "nevy".

Concrete created tranquillity by eliminating all unnecessary elements. The basic of the design concept is to create a space, which has the appearance of an old market hall. Marble was therefore a forehand choice.

The restaurant exists of two parts, the restaurant and the raw bar. To create a visual separation between the two parts black and white are chosen. In the white part, the ceiling and floor are black. The opposite appears in the black area. There you see a white floor and ceiling.

In this part of the white restaurant, you can find the best view of the "Ij" and Amsterdam north. Concrete chose to place six couches, white marble tables and white high gloss revolt chairs cross-grained the façade. In this way guests can optimally enjoy the view.

The couches have a "deeper" seating so that the customers can relax longer after finishing dinner.

In the back of the restaurant you can find the raw bar. In this area the chef is the main focus point.

The raw bar experience is different compared to the white restaurant; the walls are made of black marble, the floor of white marble and the ceiling is painted white.

混凝土设计事务所应邀对奥纳西斯餐厅进行重新设计。这家餐厅将与已有的"恩维"美食餐厅和"维聂"酒吧共同组成美食三部曲。"聂维"是一家以鱼菜为主的餐厅。

设计师去除了所有不必要的元素，为餐厅营造了平静的氛围。基本的设计概念是在空间中建立一个类似旧市场的大厅。表面以大理石为主材。

餐厅包括两部分，餐厅和生食吧。设计师以黑白两种色调从视觉上将这两部分区分开来。在白色部分中，天花板和地板是黑色的。黑色区域则相反，地板和天花板为白色。

白色的餐厅中，观赏北阿姆斯特丹美景的角度是最好的。设计师在这里放了六个沙发、大理石桌和白色的高脚椅。这样客人就能尽情享受眼前的美景了。

沙发的座位设计得更深，顾客在用餐后可以放松身体。

餐厅的后面是生食吧。在这里，厨师才是关键。

生食吧与餐厅不同，墙壁是黑色的大理石，地面是白色大理石，天花板也漆成了白色。

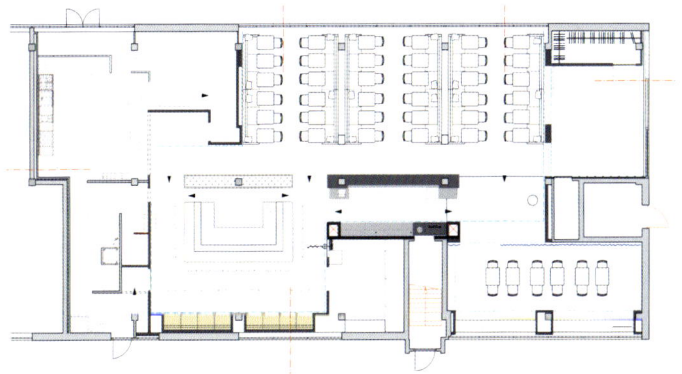

Vengeplus Capacity

闻佳餐厅

Project Name: Vengeplus Capacity **Designer:** Nagehan Acimuz **Location:** İstanbul, Turkey
Photographer: Nagehan Acimuz **Time:** 2007

项目名称：闻佳餐厅 设计师：纳格汉·阿西姆兹设计事务所 项目地点：土耳其，伊斯坦布尔
摄影：Nagehan Acimuz设计事务所 完成时间：2007年

The designers created 330 sqm indoor and 440 sqm outdoor seating area which accomodates 400 customers in total. In the whole restaurant area the designers have used stone wall coverings seperated by wooden semi-transparent seperations which were designed to both seperate different types of seating units and get a wide point of view. Floors and 13 metre bar were covered by natural stone. The lay-out plan was formed by using free tables along entrance axis, a long social table, different sizes and shapes of small tables which could be combined and seperated according to the number of customer groups.

The deisgners place most of the small tables in front of the long banquette seatings. Some compartment like private seating areas were formed between wooden seperations. Ceilings covered by wooden pannels around the area were lowered on top of the built-in seating to create a sense of coziness. Some suspended lighting fixtures were chosen to have a dim light for late night dinners. The other parts of the ceilings, on the contrary, were kept high and light-coloured to help keep the spacious feeling. Wooden seperations and wall coverings were supported by sketch images of vegetables to add a sense of humour to the whole place.

设计师分别在室内外建造了330平方米和440平方米的就座区，总共可容纳400位顾客。石制墙面用木质的半透明隔层隔开，以分离不同的座位区，扩大空间的视野。地板和13米的吧台皆由天然石质材料装饰。同时设计师采用沿入口轴线而立的随意桌、一张用来交流的长桌以及依顾客人数可拆分组合的不同规格和形状的小桌，将布局合理规划。

长软座椅前设有若干小桌，并用木质隔层打造类似私人座区的隔间。座区周围的天花板用木质面板装饰，并稍作下调以打造室内座区及顶部协调的舒适之感。室内选用吊灯为夜间的晚餐打造出朦胧氛围。而其他部分的天花板，则被拉高并用浅色打造出空间宽敞之感。木质隔层与墙围用蔬菜内容的图画装饰，更为整个空间增添些许诙谐幽默之感。

SEVVA Restaurant

SEVVA餐厅

Project Name: SEVVA Restaurant   **Designer:** TsAO & McKOWN Architects   **Location:** Hong Kong, China   **Time:** 2008

项目名称：SEVVA餐厅 设计师：曹&麦克恩建筑师事务所 项目地点：中国，香港 完成时间：2008年

A bare rooftop with superb views is transformed into multiple dining environments, each with a distinct menu and character. SEVVA was envisioned as a miniature city within a city, a composition of distinct "neighbourhoods". Each one embodies the essence of one of SEVVA's menus, thereby offering guests a panoply of experiences to choose from, much as any great city does.

Bank Side caters to Hong Kong's business and power elite, with vistas of Norman Foster's groundbreaking HSBC (Hong Kong Shanghai Banking Corporation) headquarters and the commercial district as backdrop.

Harbour Side offers a more casual dining experience; warm tones of ivory, ebony, and buttercup yellow serve as counterpoint to Bank Side's clean-lined Minimalism. A unique vaulted ceiling casts a soft, luminous glow over the room. Taste Bar transforms itself from a bento box bar at lunch time into an urbane cocktail lounge in the evening. A bold tropical palette, complete with faux rhinoceros horns and a lush vertical garden of aromatic maiden hair ferns from Belgium bring a whimsical touch of the exotic to this ultra-metropolitan setting. Cake Corner is a child's dream made real. Confections morph into décor in this delirious chamber of glass-encased treats. Surrounding it all is a wrap-around terrace: surrounded by iconic skyscrapers, harbour, and mountains close enough to touch, one feels viscerally connected with the entire metropolis at once.

SEVVA餐厅选址在一幢建筑屋顶露台上，就餐环境多样化，每一区域都具有自己的特色菜肴和风格。这里如同一个迷你城市，为顾客们提供一次全套的奢华体验。

岸边区（Bank Side）以汇丰银行总部以及著名商业区为背景，是香港商界、政界精英的首选。海港区则以随意的就餐环境为主：象牙色以及黄色等暖色调与岸边区线条明朗的简约风格形成互补；柔和的光线从拱形天花上洒落。品味吧（Taste Bar）功能多样，中午出售盒饭，晚上则转换成了鸡尾酒吧。热带风格的色调搭配人造犀牛角饰品营造了趣味十足的氛围。蛋糕角落（Cake Corner）是儿童的天地，甜点摆放在玻璃展架内，如同装饰品一般。四周被露台环绕，远处的摩天大楼、海港以及触手可及的山峰让人感觉如同置身于大都市中。

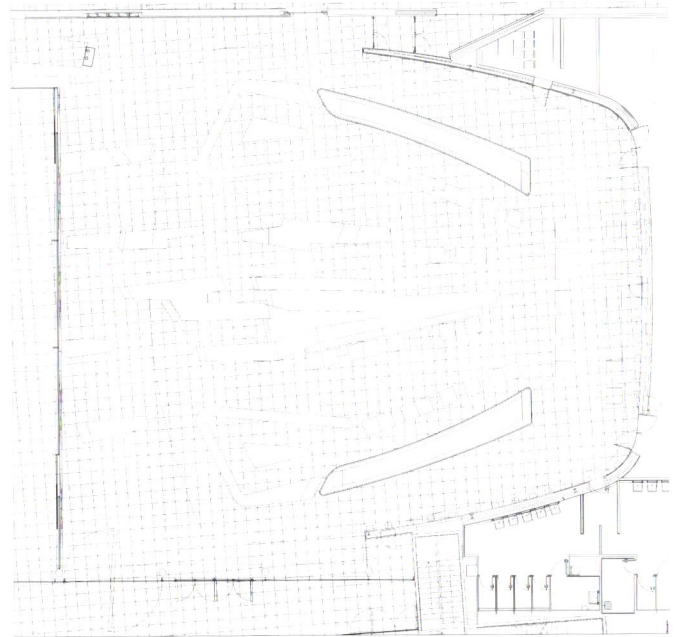

BRAND STEAKHOUSE

品牌牛排餐厅

Project Name: BRAND Steakhouse   **Designer:** GRAFT   **Location:** Las Vegas, Nevada   **Time:** 2008
Size: 7,500 sf

项目名称：品牌牛排餐厅 设计师：GRAFT设计事务所 项目地点：美国内华达州，拉斯维加斯 完成时间：2008年
项目面积：约70平方米

The BRAND Steakhouse sits amidst the casino floor in the Monte Carlo Hotel in Las Vegas. The marriage of texture and form coupled with numerous height variations affords a constantly diversified spatial experience. The perception is individualized and ever-changing due to the magnetic patterning of the ceiling panels as well as the rich and ambient materials, invoking the sensation of being present within a diversified ecology. The choice of materials — smoked oak flooring, leather, suede upholstery, tree branched mirror and LED animated panels, cow hide patterns — gives the space an essence of nature serving as a counterpoint to the hyper-synthetic casino environment.

The façade's low walls, whose distribution coheres to the ceiling panels above, resurrect the image of the wooden fence typical of a ranch. A low wall opens up to the casino and provides smooth transition into an enticing space.

The ceiling canopy could be termed as the jewel of the space. Consisting of numerous panels wrapped in custom-printed fur, the canopy reveals itself as an intricately distributed cowhide pattern, which is abstracted through the technique of pixilation. Panels lit at various depths create a spectacular dusting glow, intended to captivate the viewers and simultaneously evoke feelings of comfort, repose and visual movement. The panels gradually rise towards the back of the restaurant, unfolding the guests to the maximized view of the animated ceiling.

餐厅坐落于拉斯维加斯的蒙特卡洛酒店的俱乐部楼层，材质和结构的密切结合，以及高低层次的变化，带给顾客一种别致且变幻多彩的空间享受。其设计理念在于：对顶棚天花板进行类似磁性图般的独特设计、丰富的材料选择和包装，给顾客一种置身于变幻莫测自然环境里的感觉。材料的选择——烟熏栎木地板、皮革、绒面革摆设、树枝样的镜子，在灯光的映照下显得生机勃勃的嵌板，以及牛皮图案，都给空间增加了一种类似置身大自然的感觉，从而与快乐兴奋的人为娱乐环境形成了对应。

餐厅门前装饰有很多花墙，墙上的装饰延伸至餐厅顶部的天花板，重现了农场木栅栏的典型形象，其中的一面矮墙被打开，为联通至俱乐部——这个热闹兴奋地空间，提供了一条自然的过度带！

天花板的罩篷是整个空间最引人注目的地方——其间分布着无数的镶板，它们被嵌入按照当地习俗印花的皮毛上，使得那些镶板看起来像是杂乱分布的牛皮革样式，这种设计方法源于对实体动画技术的抽象理解！镶板在不同的深度被灯光照亮，从而形成了一束束壮观且引人入胜的射光，使顾客为之陶醉的同时，还能舒舒服服的享受其中。镶板在餐馆的后部分逐渐升高，使顾客能观察到整个天花板的顶棚。

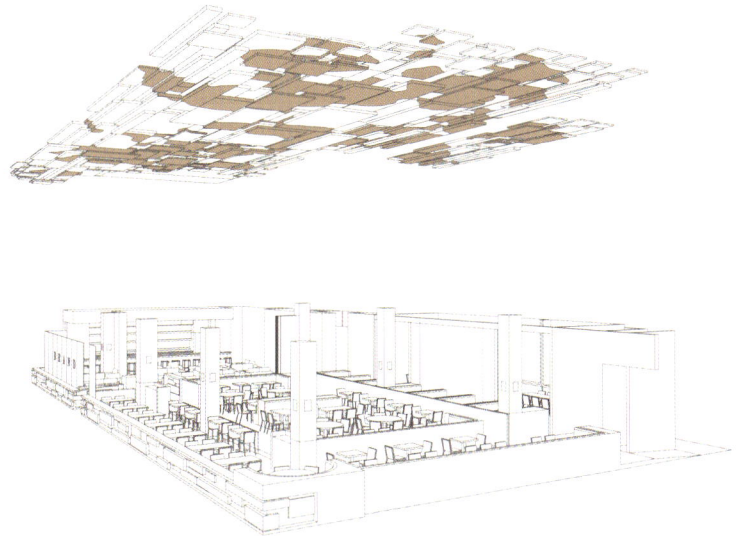

Inamo

埃那谟餐吧

Project Name: Inamo   **Designer:** Blacksheep   **Location:** London   **Photographer:** Francesca Yorke
Time: 2008

项目名称：埃那谟餐吧 设计师：Blacksheep设计公司 项目地点：英国，伦敦 摄影：弗朗西斯卡·约克 完成时间：2008年

Inamo is 310 sq m (gross), 62-cover Wardour Street restaurant and downstairs bar, offering customers high-quality Asian fusion cuisine. It also offers a whole new paradigm in the way people can order food, with menus projected onto tabletops, allowing diners to order food and beverages interactively, to change the ambiance of their individual table, to play games or even to order up local information and services, such as booking a cab.

It was an ambitious project and a tricky concept to get right. Great attention had to be paid especially to ensure that lighting levels and proportions were spot on for the restaurant to work at all hours of the day and night. The restaurant design also had to have an overall strong sense of identity as a space, neither overwhelming nor being overwhelmed by the technology at its heart.

'Cocoon' projectors are set at the same height throughout within the suspended high gloss black ceiling and come in three sizes to light 2-cover, 4-cover or 6-cover tables. When customers sit down there are white spots for plates and an individual 'e-cloth' for each table. Customers use a touch panel to order food and drink or change their table top to one of the seven other patterns available.

The restaurant has been treated partly in monochrome to allow the tabletop illuminations to stand out at its heart. And partly with further graphics (inspired by the colour choices of kaleidoscopes and origami for an Asian feel) to mirror the strong graphic sense created by the tabletops. Walls are therefore in a white vinyl wallpaper with mirrored graphic panels, which work effectively like lightboxes with an etched pattern and a cut pattern, with light allowing the cut pattern to shine through. Seats are either a silver vinyl banquette with feature red stitching or else 2-tone flip chairs with black backs and white frontage. The tables themselves have a black base and a white Corian top.

"埃那谟"餐吧是一间总面积310余平方米、坐落于沃德街的餐馆和地下酒吧，它为顾客提供高品位的亚洲美食！同时，它也为人们点餐提供了一种新的方式：把菜单投影在桌子上，用餐者可以通过桌面的触摸屏电脑预定食物和饮料，从而改变了个人餐桌的氛围，顾客还可以玩游戏，甚至定制当地信息和服务，例如：预定出租车。

这是一个充满挑战性的项目，设计师们不得不花费很多心血去确保灯光的水平位置和对称性，用以保证灯光可以24小时发挥作用。餐馆设计不得不具有一种空间感的特性：既不会被其内部的工艺压倒又不会以一种压倒一切的姿态存在！

"蚕茧"投影仪被悬浮安置在整个黑色顶棚的同一高度上，它们分为三种型号：即双人桌、4人桌和6人桌，当客人坐下的时候，便会有为碗碟准备的白点出现，并且每张桌子都会有单独的"e——屏幕"。客人不仅可以使用触摸屏预定食物和饮料，还可以在备选的七种桌面中改变他们自己的桌面！

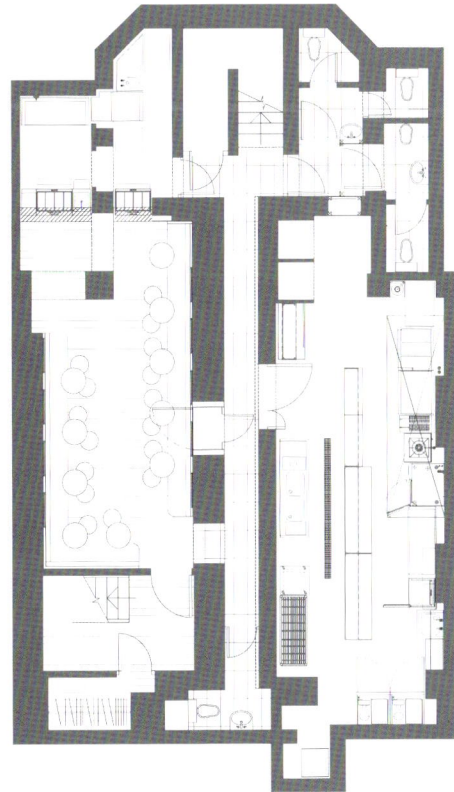

"pearls & caviar"

"珍珠和鱼子酱"

Project Name: "pearls & caviar"   **Designer:** concrete   **Location:** Arab   **Photographer:** concrete
Time: 2008

项目名称：珍珠&鱼子酱餐厅 设计师：concrete设计公司 项目地点：阿拉伯 摄影：concrete设计公司
完成时间：2008年

"pearls & caviar" represents the new Arabian lifestyle, a luxury fusion of east and west, black and white, occident and orient, light and shadow, the extrovert and the introvert, the intimacy and the view.

The basic idea of the design of the restaurant was to create an abstraction of the commonly used oriental forms and materials without loosing their richness. To achieve this, the amount and the richness of the oriental patterns and forms which are found in traditional oriental spaces are kept but without color. All colours are replaced by either shades of black or shades of white — both in combination with silver. Therefore the restaurant is divided in two parts, which also gives it's its name: "pearls & caviar".

"珍珠和鱼子酱"餐厅代表了阿拉伯人新的生活方式——东与西、黑与白、西方与东方、光与影，内外兼收、兼容并蓄的奢侈融合。餐厅最基本的设计构思即为创造一个集东方精华内容与形式的抽象环境，但同时要求不失其丰富性。为达到这一目的，设计师保留了东方特有风格及形式（在传统中式空间内随处可见），但却摒弃了色彩的运用，黑色或白色占据整个空间，餐厅也因此被分成两个部分，如同其名字一样，"珍珠"和"鱼子酱"。

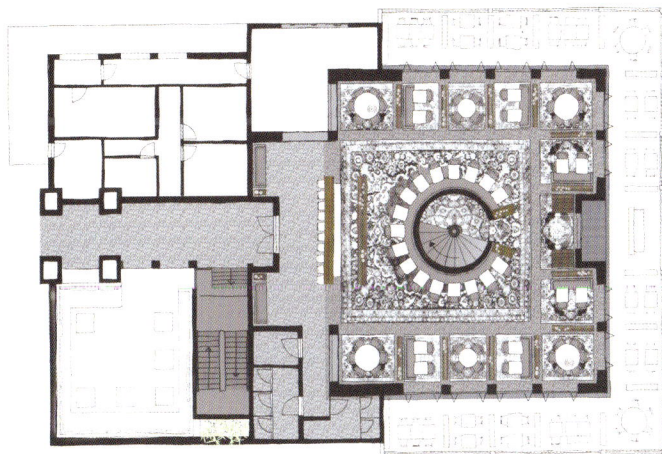

Lido

莱多餐厅

Project Name: Lido Designer: JP Concept Pte Ltd Location: Singapore
Photographer: JP Concept Pte Ltd Time: 2007

项目名称：莱多餐厅 设计师：JP Concept Pte 有限公司 项目地点：新加坡 摄影：JP Concept Pte有限公司
完成时间：2007年

Nestled in the tropical greenery of the Sentosa Golf Club, a sense of grandeur and luxury permeates this restaurant. The colour concept juxtaposes an off white colour scheme against a black background to create spatial depth and visually arresting contrast. Selected furniture are used to infuse a dash of contemporary vibe to the lavish setting. In the dining area, brown tinted mirrors on dark blackish colour timber veneer serve to provide a "scattering effect" to reflect the outdoor greenery into the room. A long suspended tinted glass top counter provides storage solutions while maintaining the elegance of the setting. Parallel to the walls are long black suede modular booth seats forming a passage between dining area and the bar. Placed opposite these long booth seats are specially selected transparent polycarbonate "Louis Ghost" chairs. The modish chairs make a perfect "transitional" element bridging and balancing the more ornate design aspects with the chic. On the white table settings in the middle of the room is the alternating orange & green "Mademoiselle" armchairs. Crowned with richly coloured upholstered top, rectangular backrests and concave sides on transparent legs, these armchairs complete the smooth blend of classy styles with a burst of contemporary chic.

The lounge on the other side is an intermingling of both bright and dark elements. The main focus is the bar counter inspired by the human curvature. The counter's customised fibreglass moulding is illuminated by built-in lighting. 5 sleek black and 1 white "Spoon" high stools add elegance to the meandering centre piece. The ZEPPELIN chandelier, is used as the central lighting in the lounge. Unlike traditional sharp edged and elaborately wired chandeliers, the cocooned chandelier presents a softer natural feel.

该项目位于圣淘沙高尔夫俱乐部,整间餐馆富丽堂皇。黑白搭配更显空间立体感,颜色的对比鲜明,给人带来视觉冲击。精挑细选的家具为环境增添了现代感。

在就餐区,深黑色薄木镶板上安置棕色彩镜,它发挥散射效应,将户外绿意反射到该区域。一个悬空的彩色玻璃面的柜台,营造优雅氛围的同时还具备存储的功能。与墙平行而设的一排黑色绒面的包厢座椅在就餐区与酒吧之间形成一个通道,而特别挑选的透明聚碳酸酯材料的精灵椅放置在对面。时尚的座椅充当了完美的过渡元素,平衡了华丽的设计。在屋子中间安放白色桌子,橙色和绿色扶手椅交叉摆放。这些扶手椅带有矩形的靠背和透明的椅腿,椅子上面铺有彩色的华丽软垫,融时尚与优雅于一体。

休息室在就餐区的对面,其中混合了明暗两种设计元素。空间的焦点便是吧台,其设计灵感源于人的曲线轮廓。定制的玻璃饰品在内置灯打出的光效烘托下格外迷人。五张亮黑色、一张亮白色高脚凳为聊天环境增加了几分典雅之美。作为酒吧中心照明的枝形吊灯,与传统的枝形吊灯不同,更能展现其柔美的自然气息。

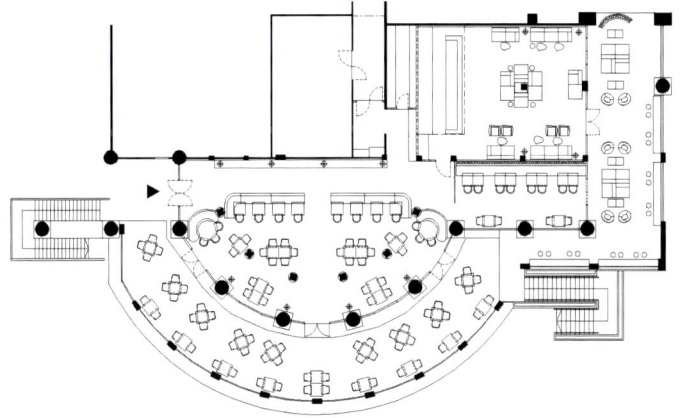

VLET Restaurant

韦丽特餐厅

Project Name: VLET Restaurant   **Designer:** JOI-Design GmbH, Hamburg   **Location:** Hamburg, Germany   **Photographer:** JOI- Design GmbH   **Time:** October 2008   **Area:** 390 sqm

项目名称：韦丽特餐厅 设计师：JOI设计有限公司 项目地点：德国，汉堡 摄影：JOI设计有限公司 完成时间：2008年 面积：390平方米

In Old High German the word "Vlet" meant a fleet, or a channel in coastal cities. One manner of introducing the context of the restaurant's name into the interior design is through the gigantic golden fish sculpture that greets guests at the entry. The space plan also capitalizes on the "stream-like" ripples created by the curves in the barrel-vaulted ceiling to create a directional flow though the dining area. Additionally, a floor lamp constructed from soft fabric to resemble a floating sea urchin is placed alongside the timber dining table by the feature window. The predominantly natural tones of the restaurant's décor allow the vibrant culinary creations to be the true artistic masterpieces. Reminiscent of the richness of the traders' treasures of coffee, carpets, spices and silks stored in these former warehouses, JOI-Design has created a canvas that allows the menu to shine.

The graceful arches of the barrel-vaulted ceiling, steel beam supports and textured brick and plaster walls are enhanced by newer design elements such as the rustic timber plank floors, rich leather chairs and driftwood "sculptures" lit from below to define their shadows against the irregular wall surface. The heavy, rough-hewn natural finishes of these structural design features are offset by the smooth, shiny modern materials of the stainless steel dining table bases and barstool frames, clear crystal wine glasses and sparkling translucent pendant lights. The visually exciting contrast is further emphasised by the bold flash of fresh lime green behind the bar and on its counter surface.

在德文中 "Vlet" 意为港湾或海峡，为在设计中体现餐厅名字的背景，设计师们专门在入口处打造金鱼雕塑，用于欢迎客人的到来。为进一步深化主题，天花板上的曲线造型同样营造了溪流般的感觉，使得空间布局格外流畅；落地灯由柔软的织物打造而成，沿着窗户旁边的餐桌设置，如同流动的海洋。

优雅的拱形屋顶、古老的钢梁支柱以及质感十足的砖石墙壁在木板地面、皮质椅子以及底部照明的木质雕塑的陪衬下更加韵味十足。粗糙自然的装饰特色与光滑的材质相得益彰，不锈钢餐桌和凳子、晶莹透明的玻璃酒杯以及闪闪发光的灯饰格外引人注目。此外，吧台后面及表面一抹淡雅清新的黄绿色调将原有的对比特色进一步深化，带来了活泼生动的气息。

Chocolate Soup Café

巧克力汤餐厅

Project Name: Chocolate Soup Café   **Designer:** pericles liatsos designers
Location: Limassol, Cyprus   **Photographer:** pericles liatsos designers   **Time:** 2008

项目名称：巧克力汤餐厅 设计师：伯里克利·利亚特索设计师事务所 项目地点：塞浦路斯 莱梅索斯
摄影：伯里克利·利亚特索设计师事务所 完成时间：2008年

Chocolate soup is a new concept cafe located in the centre of Limassol, Cyprus. The space is divided into 3 parts. The large sofa on one end, the central area with various shapes of sofas, poufs and chairs and the high tables area. The way the furniture was placed indicates a playful mood and atmosphere that makes the customer feel relaxed. The colours are simple and vibrant, with the various shaped sofas and the oversized chandelier giving it a more surreal feeling to a quite small space. Though not the standard cafe layout with large sofas and poufs combined with high tables and stools, classic armchairs and chandeliers, the visitor surely enjoys this strange composition.

The exterior canopy is see-through and of fuchsia colour which gives the whole space, exterior and interior, a different feeling depending what time of the day it is. It is decorative as well as functional. The canopy gives color to the white rattan sofas and armchairs at midday and protects from rain or humidity at night.

The space attracts customers for coffee and breakfast in the mornings, business lunches, afternoon coffees and on a smaller scale dinner and drinks. A very enjoyable and happy space, a tribute to chocolate!

巧克力汤餐厅位于梅莱索斯市中心，室内空间分为三部分：一侧摆放着长长的沙发，中央区是形状各异的椅子，另一侧则是高脚桌。独特的家具组合方式营造了趣味十足的空间氛围，让顾客置身其中备感舒适。此外，餐厅内色彩简约而不失艳丽感，让人耳目一新。

室外天篷由透明材料打造而成并饰以紫红色，装饰性十足而又不失功能性——中午赋予白色沙发和摇椅亮丽的色彩，夜晚阻止风雨进入。餐厅吸引着大量的顾客，早上前来喝杯咖啡、吃点早餐，中午和晚上尽情享受美食。这是一个欢乐的天堂！

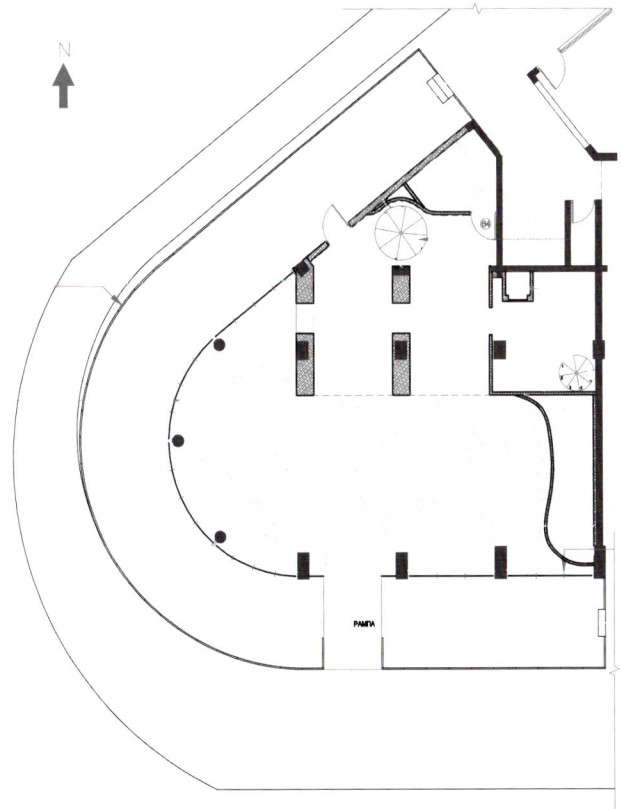

Mc Donalds Urban Living Prototype

麦当劳—城市生活设计原型

Project Name: Mc Donalds Urban Living Prototype   **Designer:** Studio Gaia, Inc
Location: New York, USA   **Photographer:**   **Time:** 2008

项目名称：麦当劳—城市生活设计原型 设计师：Gaia工作室 项目地点：美国，纽约 完成时间：2008年

Studio GAIA was commissioned by the multi-billion corporation that helped shape American culture as well as the international food marketing in designing their brand new wardrobes that is not restricted by the industry standards but to create a brand experience that reflects fresh, modern and warm interior elements. The design focuses and celebrates on the freshness of the ingredients and promotes a fresh and healthy outlook to the household brand name. Upon entering, guests are greeted with a full height custom transparent divider panels with oversized produce graphics. Red wall tiling, light and wooden tile flooring as well as vibrant colors seating help create an open but warm dining and café experience.

The main inspirations behind Urban Living design are: Inviting, Lounging, Intimate and Relaxation. There are various main design features within the Urban Living design:

a. The yellow horizontal continues "band" that wraps the interiors from ceiling to the vertical high table (TV) wall then transitioning to the floor;

b. Light wood ceiling panels with oversized pendants above "lounging banquets" creating a warm and inviting feel;

c. Custom black and white graphic wall covering and semi-see thru panels creates privacy and graphic ideas based on store localization.

GAIA工作室受邀负责麦当劳室内空间设计，要求摆脱行业标准的制约，打造一个全新的现代化氛围，旨在突出麦当劳清新健康的形象。为满足这一需求，设计师在入口处设置了与天花板等高的定制玻璃隔断板，上面绘制着店内食物的巨幅图片。红色瓷砖墙壁、木质地板，配以活力十足的颜色，营造了愉悦而温暖的就餐氛围。

"城市生活"理念由四个要素构成：热情、休闲、亲切、舒适，特色主要表现在：

a.黄色包裹着整个空间，从屋顶延续到墙壁再延伸到地面上；

b.休闲宴会厅内木质天花板以及悬挂着的巨型吊灯增添了温暖舒适的氛围；

c.定制的黑白相间墙面装饰以及半透明的板条营造了私密感。

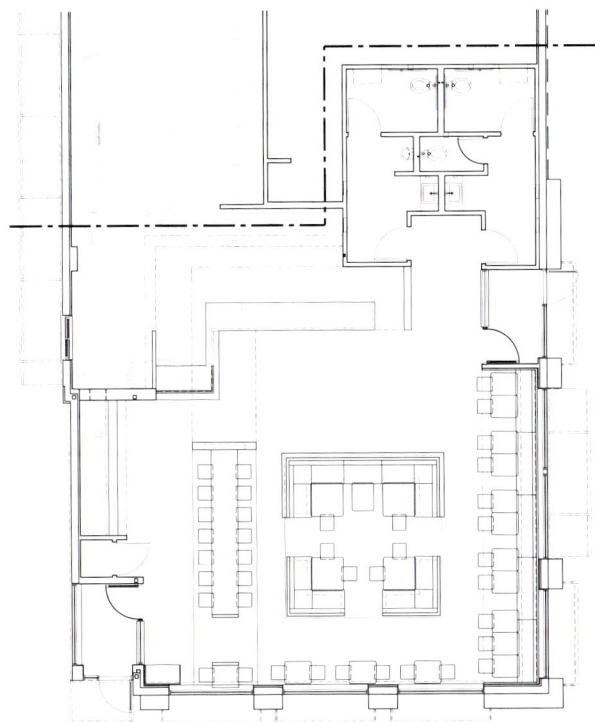

Oth Sombath Restaurant

奥斯餐厅

Project Name: Oth Sombath Restaurant   **Designer:** Jouin Manku   **Location:** Paris, France
Photographer: Eric Laignel   **Time:** 2008

项目名称：奥斯餐厅 设计师：朱伊恩·曼梧 项目地点：法国，巴黎 摄影：艾里克·莱格尼尔 完成时间：2008年

Searching for the right balance in ambience for the restaurant — it should at once respond to the restaurant's cuisine and offer something exquisitely modern. The designer immersed themselves in the universe of Thailand and translated it into the Parisian context. They studied the traditions and culture for colours, rhythms, volumes, light, while simultaneously looking to create a fully contemporary gastronomic experience. The interior decoration uses stunning elements, such as smooth lines and sleek curves, something sensual and exotic, something that feels like the delicate blend of all these rich influences embodied in one place.

The inspiration runs through the bar and three dining rooms, each offering a unique atmosphere to enjoy Oth's sumptuous flavours – one in scintillating gold, one in bold orange, and one in soft creams – the colour palette calling to mind in turn the temple treasures, bold colours of the women's dresses and rice paddies. Saffron-hued upholstered walls pop in the restaurant's multilevel interior. Textures also play reference in the restaurant – that of the curving wall of the bar looks like the hairstyles on Buddha sculptures while the warm wooden floor on the rez-de-chaussée echoes the dark wooden artisans' creations in Thailand. The arch of a dragon's back resonates with the sweeping curves of the stairway connecting the three dining rooms.

为在餐厅中打造一种平衡的氛围——与菜肴相得益彰，同时彰显现代风格，设计师充分运用泰式特色，并将其成功移植到巴黎特有的环境背景中。室内设计特别选用令人眼前一亮的元素，如平滑的线条和优美的曲线，营造了别有韵味的感官享受。

设计师的独特灵感在酒吧和三间就餐室内完美展现，赋予各自不同的空间氛围。就餐室风格各异，一间饰以闪亮的金色，一间以橙色为主，另外一间则完全采用柔和的奶油色。织物同样被用作主要设计元素，装饰在酒吧的墙壁上使其看起来如同大佛雕塑的头饰。楼梯蜿蜒而又壮观，将就餐区连通。

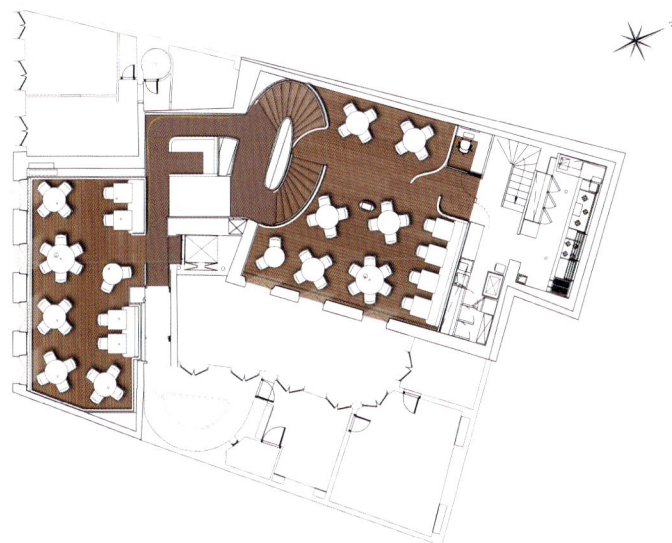

Restaurant Alain Ducasse

艾伦—杜卡斯餐厅

Project Name: Restaurant Alain Ducasse　**Designer:** : Patrick Jouin　**Location:** Dorchester UK
Photographer: Eric Laignel　**Time:** 2007

项目名称：艾伦—杜卡斯餐厅 设计师：帕特里克·茹安 项目地点：英国，多尔切斯特 摄影：埃里克·莱涅尔
完成时间：2007年

Concept

The restaurant at the Dorchester is a presentation of English culture as digested by the French. The project takes inspiration from the ever-present power of English gardens, parks as well as horse races, horse riding, picnics, classic cars, the country house, and a sprinkling of Midsummer's Night Dream. The aim of the project is to find a way to create a luxurious experience that takes clients out of the city and into a new imaginative locale – a place they may recognize from books or pictures they've seen, but at the same time is something they've never really seen before. Here they seek to create a luxurious experience from seemingly ordinary materials, crafting luxury not from gold and crystal but from the particular way they combine ordinary and sometimes coarse materials, the way they curve and converge, the way the colours blend into each other and the light brings in a soft accent. Luxury emerges as more about how things are put together – a particular situation created for this meal – rather than what is put together. They can give rise to luxury, to comfort, to lavishness, simply by creating an exquisitely-considered situation.

Spaces

The entrance to the restaurant is framed by greeter's desk and menu board which give way to a small corridor. The entry corridor is dim, but it is flanked by large oval windows that give partial glimpses into the room beyond, filtered by hovering screens of shimmering green dots which in turn allow a look into the park beyond. At the end of the corridor lies a large suspended curved disk. This screen of sorts bends gently in both directions and is wrapped in leather with intricate stitching details and an inlaid luminosity. A quiet whisper, a small phrase to gently welcome the guests to the restaurant as well as separate them from the rest of their day, evening, night...

The main dining room is filled with light. The walls are clad in seemingly traditional raw oak woodwork with a floor of wide oak planks. The ceiling caps the room in plaster. To one end of the room lies a series of bent screens.

设计理念

艾伦—杜卡斯餐厅可以说是法国人对于英国文化的诠释的展示，从花园、公园、赛马、骑马、野餐以及《仲夏夜之梦》等英式特色及传统中获得灵感，旨在以一种新的方法为前来就餐的客人营造奢华体验——一个他们在书中或画面中似曾看到但却没有真正去过的地方。他们摒弃了金子以及水晶等贵重元素，大量运用普通的材质和色调——以特殊的方式将其组合，打造一种令人惊讶的奢华格调，同时兼顾舒适。

空间

主入口处是接待台及菜单板。一条狭小的走廊内光线昏暗，开阔的椭圆形大窗悬在两侧，视线穿过这里就可以"窥见"两侧的就餐室内，装饰着绿色小点的"悬浮"屏风格外吸引眼球。走廊的尽头悬垂着一个巨大的弧形桌子。

主餐厅内光线充裕，墙壁采用传统风格的橡木结构饰面，工艺感十足。地面同样采用橡木板铺设，天花则用石膏装饰。房间一侧的尽头摆放着一系列的弯曲的屏风。

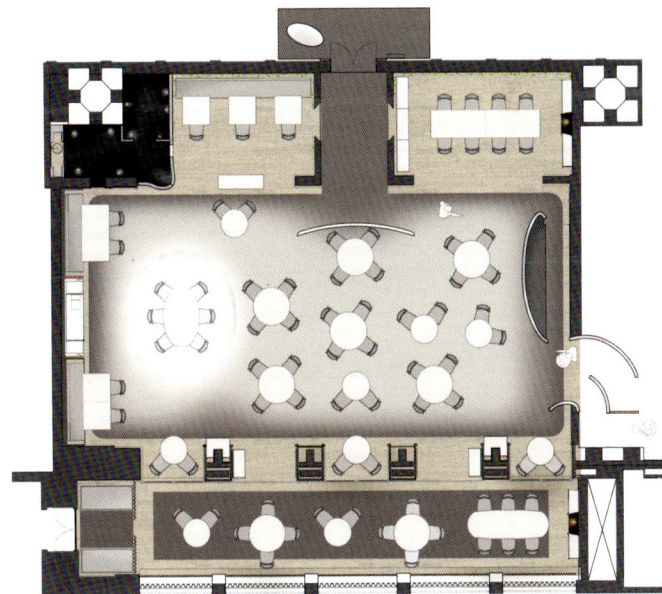

Leggenda Ice Cream and Yogurt

莱真达冰激凌乳酪店

Project Name: Leggenda Ice Cream and Yogurt **Designer:** SO Architecture
Location: Ramat Ishay Israel **Photographer:** Asaf • Oren **Time:** 2009

项目名称：莱真达冰激凌乳酪店 设计师：SO建筑事务所 项目地点：以色列，拉马特 摄影：阿萨夫·奥伦 完成时间：2009年

In the industrial zone of Ramat Ishay northern Israel, inside an old abandoned carpentry shop a new branch of the Leggenda ice cream and yogurt was put up. It wasn't easy to design a space with an airy and calm feeling, as the customer desired, in the space the deisgners had to work with. This difficulty was caused by very low ceiling of only 2.30m, the very long and narrow proportion of 4.50m X 18.50m of the place. Another major difficulty was caused by the foundations of the building above the shop, which invaded into the inner space.

The main idea of the design was to create two linear elements that will lead the client into the deepest parts of the shop where the cozy sitting places are located. The large scale window in the entrance and on the side of the hall was opened to allow daylight in. The entrance is emphasised by its geometry. The large door and facade were slightly submerged into the inner space in order to entice the visitor into the place. The steel door and welcome roof were created as a single piece. The graphic wall behind the goods displays was designed by FIRMA, and adds fun to the atmosphere of the shop.

莱真达冰激凌乳酪店位于拉马特工业区一幢废弃的木艺工厂内，原有空间屋顶高度仅为2.3米。设计师面临的挑战是如何打造一个通透而淡雅的环境。

基于客户的要求，设计师打造两个线性结构，便于指引顾客走到最里面的座区。入口处设计强调几何造型，宽阔的大门一直通往室内。为使得室内光线充足，设计师在入口以及大厅的一侧开启了宽敞的大窗。商品展示架后面的墙壁是专门设计的，上面画满了各种图形，为整个商店带来一丝童趣。

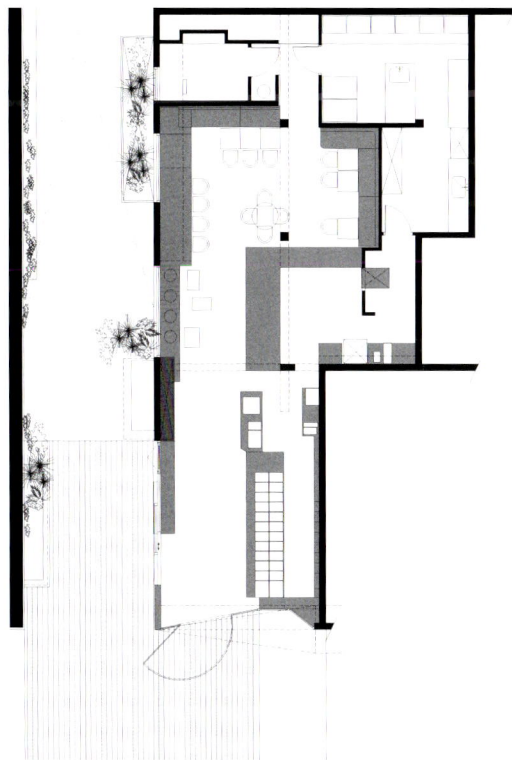

Rosso Restaurant

罗索餐厅

Project Name: Rosso Restaurant   **Designer:** So Architecture   **Location:** Ramat Israel
Photographer: Asaf Oren   **Time:** 2007

项目名称：罗索餐厅 设计师：So建筑公司 项目地点：以色列，拉马特 摄影：阿萨夫·奥伦 完成时间：2007年

The Rosso restaurant is situated in Ramat Yishay, a town in the valley of Yezre-el in the Northern part of Israel.

The green hill around the place influenced the design. The planning seeks to explore – How to give to the restaurant space the feeling of its surrounding. The design checks in what manner could it produce an atmosphere, or feeling, of being outside while sitting inside.

The atmosphere in the restaurant is pleasant. The furrowed fields, which are framed by the strip of windows, find its continuation in the ceiling, and embrace the diner as if by the foliage of trees, with the furrows flowing to the horizon.

The walls are covered in softly textured upholstery – dark gray near to the view, and a light gray at the interior walls – to enlarge the feeling of space. The ceiling is built of thirty-six steel contoured structures, and covered by painted wood. The chairs and the wide windows are decorated in plant motives to enforce the connection with the surrounding scenery. When night falls, the lighting, which is between the ceiling beams, forms a warm and somewhat mysterious atmosphere.

罗索餐厅位于以色列北部地区的小镇内，四周青山环绕，景致宜人。设计师试图寻求一种方式，使得客人坐在餐厅内，却能感到如同置身室外一般。

整体环境舒适。窗外蜿蜒起伏的田垄形状在室内得以继续——天花板模仿其设计，别具特色。墙壁采用质感柔软的材质覆面，淡灰的色调在视觉上增添了空间的开阔感。椅子和窗户采用植物图案装饰，同室外的精致相互呼应。夜晚来临的时候，灯光在天花板的梁柱之间闪耀着，营造了温馨而神秘的氛围。

VYTA – Boulangerie

维塔面包店

Project Name: VYTA – Boulangerie **Designer:** Colli+Galliano Architetti **Location:** Rome, Italy
Photographer: Matteo Piazza **Area:** 150 sqm

项目名称：维塔面包店 设计师：科利+加利亚诺建筑公司 项目地点：意大利，罗马 摄影：马泰奥·皮亚扎 面积：150平方米

This bakery was created inside a railway station, the modern symbol of speed and progress. The precise desire for integration and dialogue with the architectonic context is obvious from the very beginning. It shows that Vyta is totally projected towards the outside, and separated from it by a simple black portal. This was the theme of the interior design for the bakery, which starts from the traditional image of the bread shop; this is then reduced progressively to its essential features, in the quest for immediate and universal language that can guarantee the maximum usability in this place of transition.

The familiar colours of oak wood harmonize with the pale colours of the flooring and the ceiling, leaving space for monolithic furnishing accessories in black Corian that catalyze the attention through their symbolic purity. The polyfunctional counter is the heartbeat of every busy shop and in this case is used as an essential exhibition structure. There is a large decorative wall consisting of a paste creation symbolizing the surface and the fragrance of a bread crust.

In a delicate combination of prospects, the counter and the wall behind it curve towards the more intimate portion of the room, animated by tables and chairs submerged in a welcoming atmosphere and display alcoves similar to compact black boxes. There is also maximum design uniformity that slides unobtrusively into the busy station, while allocating ample room for the people waiting to depart, creating a relaxing ambience closely linked to the traditions that are hidden behind the minimalist lines — making Vyta a valid alternative to the world of fast food, and promoting good food culture.

维塔面包店坐落在火车站内，设计理念即为与周围的建筑氛围相一致。店内空间完全朝向室外，黑色门廊又使其与外界完全隔离。室内设计从面包店的古老形象出发，但主旨是寻求一种能够最大限度运用空间建筑语言。

橡木的色彩与地面及天花板上的灰白色相互呼应，同时为黑色可丽耐装饰营造了纯朴的背景。多功能的柜台是店内的中心元素，用于展示各种商品。高大的装饰墙壁上摆放着面团制作的各种结构，用于象征面包的光滑表面与淡雅的香味。

此外，柜台同后面的墙壁蜿蜒着朝向店内最为私密的空间，在固定摆放的桌子椅子的衬托下更加显得动感十足。这里还为那些即将启程的旅客们提供了足够的休息空间，同时借以宣传美食文化。

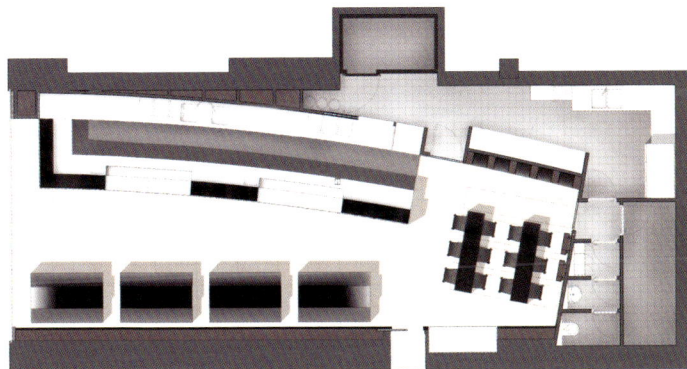

EL JAPONEZ

EL JAPONEZ餐厅

Project Name: EL JAPONEZ   **Designer:** Cheremserran   **Location:** Osaka, Japan
Photographer: Cheremserran   **Time:** 2008

项目名称：EL JAPONEZ餐厅 设计师：Cheremserran 设计师事务所 项目地点：日本，大阪 摄影：Cheremserran 设计事务所
完成时间：2008年

The floor of the restaurant is covered with a plastic carpet that evokes people of the tatami in Japanese architecture. The scarcity of columns is evident: only one column is clearly present in terms of space, which creates the impression that this long stretch is supported by only one structural element. There are other seven elements that playfully pretend to be columns but never touch the ground: they emerge from the soffit and have a specific role. The presence of wood as a material is not limited to the foreseeable use of floorboards, but contributes to the game of shapes and textures through the use of 10x10cm stud sections, which cover the solid sections and the soffit.

Over the bar, the stud wall shows some cavities, which are illuminated, making it appear less heavy and revealing that something happens behind them. A staircase hidden behind the bar leads to the rest rooms, which feature an opaque glass box contained in another wooden box. There, the environment is milder, and it playfully pretends to minimize the separation between the men's and women's rest rooms.

餐厅地面上铺设着塑料地毯，如同日式传统建筑中榻榻米。设计师采用非传统的支撑结构，它们从拱腹中延伸下来，但并不接触地面，巧妙的设计带来十足的趣味性。10厘米长的立方体小木块堆砌成形状各异的小物体，打破了木材的传统用途，同时格外吸引眼球。

吧台上方的墙壁上专门设计了一些小孔，在灯光的照射下，增添了轻盈感和神秘性。后面的楼梯一直通往休息室，木质盒子内隐藏着不透明的玻璃盒子，独特的结构成为空间的焦点。

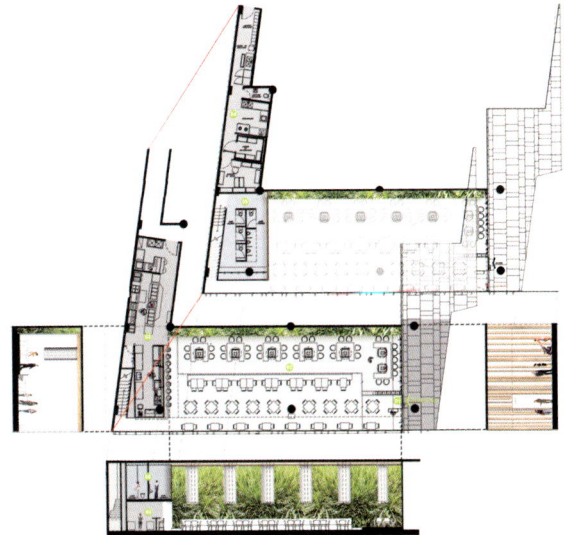

Charcoal BBQ 3692

3692烤肉餐厅

Project Name: Charcoal BBQ 3692   **Designer:** studiovase   **Location:** Seoul, Korea
Photographer: Park, Woo jin   **Time:** 2007   **Area :** 248.7sqm

项目名称：3692烤肉餐厅 设计师：vase工作室 项目地点：韩国，首尔 摄影：Park, Woo jin 完成时间：2007年
面积：248.7平方米

The designer has created the entire space with a keyword of "Layer originated from the Layer" of pork.

"Layer" means the section patterns of fatback. The fatback of pork most of Korean like has a layer of fat and a layer of meat again and again. This space shows that the layers are overlapped in many parts. Accordingly the designer has applied and repeated the "Layers" in this space. The shape of the furniture and lightings were designed one by one to express the concept. So are the repeated wires of the lighting and vertical and horizontal stripes of the furniture.

The designer has formed the layers by either using the grain of wood or piling up corrugated cardboard. The space is largely divided into the counter and waiting room, hall and kitchen. What is peculiar about this space is that the kitchen is so large to the extent that it occupies a third of the entire space. The space is warm, plain and subtle on the whole. Its feeling is a bit different and it breaks away from the typical image of the conventional restaurant. The designer painted pig-shaped objects and images, adding a sense of interest to the space.

设计师从"分层"（来源于猪肉的分层结构）一词出发，将纹理清晰的木材或层次分明的纸板叠加在一起，便于营造层次感。餐厅大体分为柜台、休息室、大厅及厨房，其中厨房格外宽敞，占据了空间的三分之一。这里有别于传统的餐厅，设计师绘制的猪一样的物体和图片使得空间内趣味十足。此外，家具和照明设备的设计同样延续"分层"的理念。

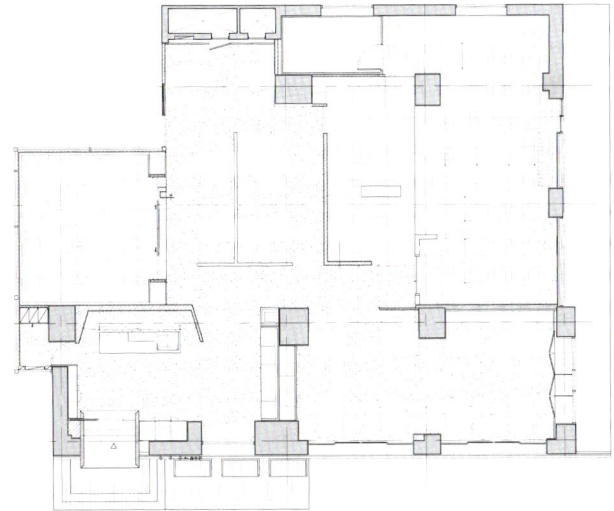

rcoal BBQ
369
LOIN
NECK BONE
TENDERLOIN
SKIRT MEAT
PORK SKIRT MEAT
HAM
JOWL MEAT
SINGLE RIBBED
PICNIC

Le Square Restaurant

广场餐厅

Project Name: Le Square Restaurant   **Designer:** MOBIL M   **Location:** Nantes, France
Photographer: Florent Degourc   **Time:** 2006

项目名称：广场餐厅 设计师：Mobil M建筑公司 项目地点：法国，南特 摄影：弗洛朗·迪格尔克 完成时间：2006年

Once upon a time there were two chefs who were also brothers. After opening two restaurants in Nantes, they decided to create a third one, close to the business quarters, therefore targeting a demanding clientele who often appreciates to be served in a rush but who still expects subtle and striking flavours as well as a well balanced wine list.

The Square offers indoor as well as outdoor meals. The outdoor space is also where ancient species of vegetables are grown for the restaurant's culinary purposes.

All the dishes are based on traditional French dishes, cooked with fresh ingredients and then enhanced with subtle flavours. In order to fulfil such a brief, a theatrical design approach was necessary: dishes are ordered, cooked and sent to the tables through an open counter that links the restaurant room and the kitchen. Pleasant cooking smells are hovering and open the appetite.

广场餐厅由两位同为厨师的兄弟创建，之前他们曾在南特市成功开设两家餐厅。临近商业区的独特位置决定了其客户源主要为行色匆匆的办公人士们，他们希望食物以最快的速度供应，当然更需要美味和调好的酒品。

餐厅包括室内、外就餐区，室外还用于种植传统的蔬菜。提供的食物以传统的法式菜肴为基础，新鲜食材为原料，巧妙地配以各种风味。为实现这一理念，设计师专门打造了开放式餐台，将就餐区和厨房连结起来。

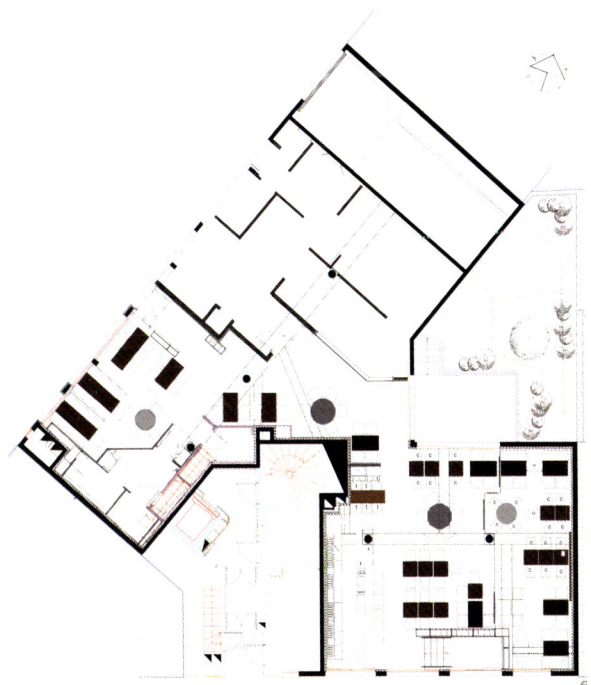

Sky 21

天空21号

Project Name: Sky 21   **Designer:** Danny Cheng Interiors Ltd   **Location:** Macau China
Photographer: Danny Cheng Interiors Ltd   **Time:** 2007

项目名称：天空21号 设计师：郑炳坤室内设计公司 项目地点：中国，澳门 摄影：郑炳坤室内设计公司 完成时间：2007年

This project brings a new challenge to the designer, to combine a bar, a dance club and a restaurant together in one place. Sky 21 is contemporary, minimal space with architectural forms combined.

On the 21st floor, a restaurant with a sushi bar is located at one end of the room. Semi-circular red sofas are cosy and better than having standard chairs and tables. Upon arriving on the 21st floor customers are welcomed by a black glass panel water feature stretching two floors. An outstanding effect is evinced when the adjacent colour changing LED-lit staircase connecting to the other floors like 22nd floor is seen through the water feature. The zone around the restaurant's sushi bar is capped with a low-lying black ceiling with exposed piping and lined with timber floors and black granite walls and mirrors and is separated from the VIP rooms with a UMU electrical semi-glass sliding door.

Upstairs, 22nd floor is designed as a two-zone bar, a separate lounge area and disco areas. This floor turns into a disco place filled with hopping dance music after 10 pm and different moment is functioned differently. On the 23rd floor, the roof, is a rest place for those who wish to take a break from pounding music and dancing. Placed with daybeds on a wooden platform, one feels like staying inside a relaxing lounge. A spacious outdoor wood terrace let customers enjoy the scenery of Macau, including the iconic Macau Tower and the ocean. There is the LED-lit dance floor with a commanding 9-LCD-panel graphic wall, the customer can relax and dance here. An open bar counter and a bed-seat area on an open wood deck is waiting them.

天空21号集酒吧、餐厅及舞厅于一身，融合多种建筑形式，风格简约但不失现代韵味。

餐厅位于21层，包括一间寿司吧。黑色的玻璃板欢迎客人的到来，透过这里便可看到装有LED照明设施的楼梯，营造了非凡的视觉效果。半圆形的红色沙发远比标准的桌椅结构更引人注目。寿司吧内，黑的的低矮屋顶与木质地板及黑色墙壁相得益彰，电动半玻璃材质拉门将其与VIP就餐区分离。

酒吧位于22层，10点之后这里会被用作迪斯科舞厅，其氛围当然也会随之转换。

23层用作休息区，为那些暂时想要逃避热烈舞曲的人们提供一个恬静的小憩之所。木质平台上放置着坐卧两用的沙发，让人备感放松。宽阔的室外露台，让人一览澳门的美丽景致。

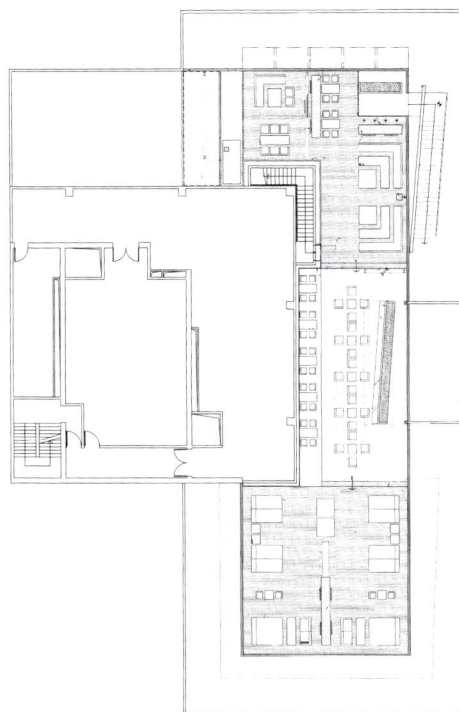

Kush 222

库什222号

Project Name: Kush 222   **Designer:** Johnny Wong and Miho Hirabayashi   **Location:** : Hong Kong Central,China   **Photographer:** FAK3   **Time:** 2008

项目名称：库什222号 设计师：强尼·黄，美穗平林 项目地点：中国，香港中环 摄影：FAK3 完成时间：2008年

222 Hollywood Road is intended to be the flagship property for Kush and offers a full time concierge in the lobby. The entry signage is finished in black mirrored stainless steel and features LED lighting. With its 45 units, the open plan apartments each average 650 square feet. Designed with either a black and white or neutral colour scheme, they all include ebony hue hardwood flooring, Light Fantastic coloured mood lighting in a shelving niche behind the queen size bed and marble-clad bathrooms.

FAK3 designed a number of custom furnishing specifically for the site. Concealed within a cabinet are a dining table and two chairs that neatly fit together and can be whisked out for dinners at home. Black Nature is a custom floral arrangement or styling element that adds an ethereal focal point to the space. The China Dreams series of accessories including the gold magazine rack and table lamp are highlights in the living room. Meanwhile the bathroom includes an oversized sliding door that allows the entire bathroom to be completely open to the adjacent bedroom, exposing the shower within a glass box for added viewing pleasure. As its crowning glory, 222's roof features a lush garden with a Light Fantastic fireplace to take away the chill on romantic winter nights.

好莱坞路222号有望成为库什的旗帜产业。这里的大堂能提供全天候的礼宾服务。入口处的招牌是黑色镜面不锈钢材质，上面饰有LED灯。这里共有45间公寓，平均面积60.39平方米。公寓内以黑色、白色和其他中性色为基调，采用乌檀木制的硬木地板，卧室大床的床头上方搭着壁龛，里面嵌着彩色射灯，浴室由大理石铺就而成。

FAK3公司为这些公寓特别设计了家具。公寓的内室里摆放着一张餐桌和两把椅子，二者搭配得如此合衬，住客可以将晚餐打包回家享用。"自然黑"系列中有造型别致的植物架和其他个性元素，把空间衬托得更加精致。"中国梦"系列的特色是客厅中夺目的金色杂志架和台灯。同时，卫生间里超大的滑动门让浴室与相邻的卧室之间几乎毫无遮挡，淋浴间也是玻璃的，看上去赏心悦目。222号最引以为傲的是它的屋顶，这里有郁郁葱葱的植物，还有壁炉为你在浪漫的冬夜驱走寒意。

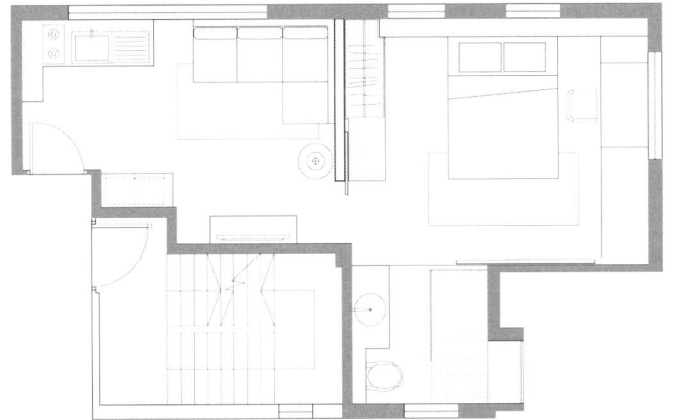

Chambers MN

明尼酒店

Project Name: Chambers MN   **Designer:** Rockwell group   **Location:** Minneapolis , US
Photographer: Stuart Lorenz   **Time:** 2006

项目名称：明尼酒店 设计师：罗克韦尔设计组 项目地点：美国，明尼阿波利斯市 摄影：斯图尔特·劳伦斯 完成时间：2006年

The concept for Chambers Minneapolis is a spacious residence for a collector of contemporary art. Cor-ten and blackened steel as well as sand-blasted stone clad the façade, while the interior features gallery-like white walls and floors of white terrazzo and ebonized wood planks.

The hotel has sixty guestrooms and special suites, including the "Rock Star Suite". Each room and suite has a minimalist, modern feeling: an open plan layout showcases carefully curated artwork, and a sleek bathroom includes an oversized shower and extensive counter-space. On the first floor, an intimate lobby-level lounge doubles as a breakfast room and a seating area. The first-floor public spaces feature contemporary artists' artwork.

The ground-level restaurant uses an industrial aesthetic: the stair is made of blackened steel and ebonized wood treads; original stone walls have been sandblasted and left natural; and custom lighting design reinforces the industrial motif. During the day, the restaurant is illuminated by a dramatic skylight. Chambers Kitchen also services the "outdoor living room" in the courtyard between the two wings of the hotel. Here, a giant gas fed fireplace is the focal point. At night, a large-scale video installation is projected on the building façade.

明尼阿波利斯酒店的设计目标是打造一个如同现代艺术博物馆一样的宽敞的居住环境。酒店外观由考登钢、黑钢和砂岩构成，内部有画廊般的白色墙壁，地板为白色水磨石和乌木材质。

酒店共有60间客房和套房，其中包括"摇滚明星套房"。每个房间都现代前卫。房间内有开放式的艺术品展示空间。浴室的风格十分时尚，内有大型的淋浴装置和宽大的浴室柜。二楼的大厅温暖舒适，并且有着双重作用，既可作为早餐厅，又是休息区。同一楼层的公共区挂满了当代艺术家的作品。

一楼餐厅充满了工业美感：楼梯由黑钢骨架和乌木踏板组成，经过喷沙的石墙更贴近自然的感觉，灯具是专门定制的，强调了工业感。整个白天，餐厅都沐浴在梦幻般的自然光之中。厨房位于宾馆中央的院子里，这里也是"户外起居室"。巨大的天然气壁炉十分醒目。晚上，安装在酒店楼体外的大屏幕就会开始播放影片。

Carbon Hotel

凯博恩酒店

Project Name: Carbon Hotel **Designer:** PCP Architecture **Location:** Brussels, Belgium
Photographer: Sita Tadema **Time:** 2008 **Area:** 812sqm

项目名称：凯博恩酒店 设计师：PCP建筑设计公司 项目地点：比利时，布鲁塞尔 摄影：塔德玛 完成时间：2008年
面积：812平方米

The newly built Carbon Hotel and Wellness Experience is part of the vibrant rejuvenation of the commercial centre of Genk, the heart of the 19th-century Limburg coal-mining industry. The Carbon dedicates itself to life's vital element in both name and design.

The architecture office PCP developed both the building and the interiors, choosing materials based on carbon and the five basic elements – wood, fire, earth, metal and water. The facade is a succinct reference to the mining fields: jutting beyond the ground floor's wall of glass is a solid front of dark brown bricks that glisten in the sunlight. Creating another striking effect is the windows' asymmetric pattern: small square panes are recessed, while thin rectangular sections protrude as if clinging by magnetic force to the brick face. Interior walls have high-quality paints and wallpapers, which alternate matt and gloss finishes for a subtle contrast of texture and light.

Whether in the lobby lounge or in each of the 60 rooms, furnishings were selected for their honest use of materials: metal tables, wooden chairs, solid trunks as occasional tables, and fragments of stone. The use of indirect, coloured light in the rooms allows for easy adjustments to suit one's mood.

新建成的凯博恩酒店是充满生机活力的Genk商业中心的一部分，Genk商业区是十九世纪林堡煤矿产业的中心。无论是其名字还是设计，凯博恩酒店都体现了其独具特色的朝气蓬勃的理念。

PCP建筑设计公司负责整个项目的建筑和室内装饰，材料的选择主要是基于碳以及五种基本的元素——木、火、土、金和水。正面对其所处的采矿区做出了明确地暗示：在玻璃墙的旁边放置一个咖啡色墙砖，在太阳的照射下晶莹闪烁。而另一种引人注目的效果来自于窗户的不对称图样：凹陷的小玻璃窗，突起的矩形截面紧贴封面。室内的墙面采用高质量的涂料，这使得光泽度暗淡的表面同手感和亮度产生了细微的反差。

不论是在大厅的休息室还是在60个房间中的任意一个房间，家具材料都是为顾客精心挑选的：金属桌子，木质的椅子，树干做成的茶几，石头碎片等。室内灯光的色彩和朦胧感，可以很容易适应人们情绪的变化。

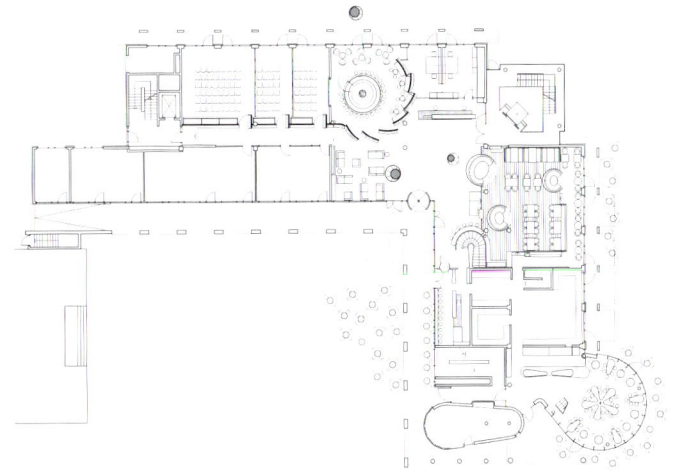

Klaus K Hotel

克劳斯K酒店

Project Name: Klaus K Hotel   **Designer:** SARC Stylt   **Location:** : Helsinki，Finland
Photographer: SARC Stylt   **Time:** 2006   **Area:**1200 sqm

项目名称：克劳斯K酒店 设计师：斯带特设计公司 项目地点：芬兰，赫尔辛基 摄影：斯带特设计公司
完成时间：2006年 面积：1200平方米

Inspired by the emotional contrasts of Finland's national epic, its nature and drama, Helsinki's Klaus K Hotel arrives bearing the stamp of Finland's finest architectural and literary traditions.

The Klaus K Hotel has transformed city landmark, formerly known for 65 years as the Klaus Kurki Hotel, with the help of major Finnish architects SARC Group and interior design by Stylt Trampoli. Located in the late 19th–century Rake building, the hotel brings Finland's national epic, the Kalevala, down to an intimate human scale throughout. The lead architects have created some of Finland's foremost modern architectural projects. Stylt Trampoli, led by Erik Nissen Johannsen, formulated the modern, contemporary inspiration of the epic implementing his craft into the hotel and restaurant design.

Each of the 137 guestrooms is given a theme illustrating the Kalevala's core emotional elements: Passion, Mystical, Desire and Envy. Klaus K aspires to go further rand "take the hotel out of the hotel": with the reception in the center of the lobby to tantalize and allure guests to visit its various food and drink atmospheres. Klaus K has created an ultradesigned life style experience where contrasts abound – Playfully delivering a luxurious experience of tradition and cutting-edge Nordic modernity.

赫尔辛基的克劳斯K酒店是芬兰最优秀的建筑和文学传统的标志,其设计灵感源于其对芬兰民族史诗的情感对比,本质与戏剧性的参照。

克劳斯K酒店转换了这个城市地标,原来由芬兰南亚建筑集团和斯带特室内设计公司设计的克劳斯酷瑞酒店曾经享誉65年,作为改成的坐标而存在。作为19世纪后期的建筑,K酒店记载了芬兰的民族强大历史,同时也记录了个人的生活轨迹。设计师曾经多次参与芬兰最著名的现代建筑项目的设计。艾瑞克尼森作为斯带特设计公司的领导者,将现代时尚的元素融入到酒店和餐厅的设计当中。

137间客房的每一间都展现着芬兰史诗核心的主题:激情,神秘,欲望和嫉妒。打造一流酒店是他们的主旨。大堂接待中心的设计,充满着魅力和诱惑,吸引着客人进入品尝各种食物和饮品。这个酒店营造了一种超现实的生活体验方式,在这里可以尽享传统的奢华和北欧的现代化。

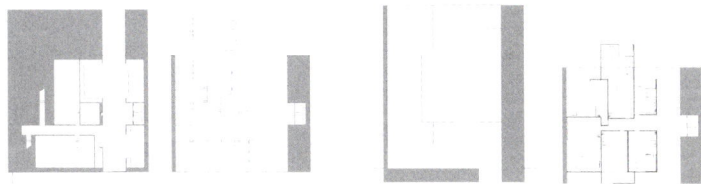

The Dominican Hotel

多米尼加酒店

Project Name: The Dominican Hotel   **Designer:** Lens Ass Architects   **Location:** Brussels, Belgium
Photographer: Lens Ass Architects   **Time:** 2007   **Area:**1000 sqm

项目名称：多米尼加酒店 设计师：水晶艾斯设计有限公司 项目地点：比利时，布鲁塞尔 摄影：水晶艾斯设计有限公司
完成时间：2007年 面积：1000平方米

Tucked behind Brussels' famous theatre and opera house, The Dominican is a new hotel that offers a strong sense of history mixed with forward-thinking, eclectic design in the European Union's capital city. The hotel's sweeping archways are a clear reference to a Dominican abbey that stood on this site in the 15th century. Its original facade has been integrated into the new construction by the Belgian Firm Lens Ass Architects. Guests enter the lofty, high-ceilinged public spaces and their breath is taken away by the attention to detail. A stroll through the Monastery Corridor evokes an almost medieval feeling of elegance with original Belgian stone flooring. The Grand Lounge, considered the heart of the hotel, calls to mind the spirit of old European decadence in soaring windows and metalwork at the same time as attracting a definitively style-conscious modern clientele with its cutting-edge design.
The 150 guestrooms and suites – each with an individual look – are situated around a quiet inner courtyard and feature a rich combination of contemporary design and luxurious textiles, offering quiet comfort in a cloister-like setting. It's a wonderful respite from the Continent's new governmental hub, and a space in which old and new Europe effortlessly melds.

坐落于布鲁塞尔著名的剧院和歌剧院的后身，多米尼加是一个崭新的酒店，营造了很强的历史氛围，同前瞻性的思想和欧洲联盟不拘一格的设计融合在一起。酒店的宽大的拱门可以让人们很自然的联想到十五世纪位于这里的多米尼加修道院。在比利时水晶艾斯设计公司的精心策划下独具匠心，它原来的门面已经被嵌入到新的建筑之中。客人们来到举架高大宽敞的公共空间，这里的细节设计令人们叹为观止，留有原始比利时石阶的地板向人们彰显着中世纪的典雅风格。奢华的休息大厅，可谓是酒店的中心所在，高悬的宽大窗户和金属摆设令人回想起古代欧洲颓废的奢华，最终奠定了这种边缘设计的地位。
150个房间和套房，各具特色——分布在一个幽静的庭院之中，展现了现代时尚设计同豪华纺织饰品的完美融合，营造了一种舒适的氛围。这是一个世外桃源，是欧洲古典于现代元素完美结合的产物。

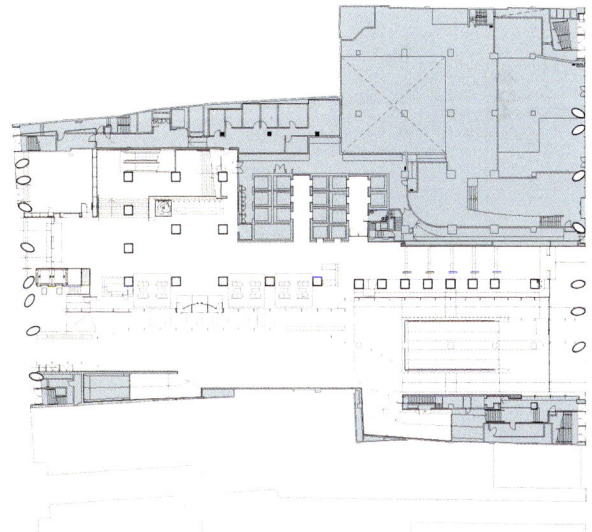

Jeronimos Hotel

杰尼贸酒店

Project Name: Jeronimos Hotel   **Designer:** Capinha Lopes& Associates   **Location:** Lisbon, Portugal   **Photographer:** Sofia Tavares   **Time:** 2007   **Area:** 1200 sqm

项目名称：杰尼贸酒店 设计师：凯品哈设计有限公司 项目地点：葡萄牙，里斯本 摄影：索菲亚 完成时间：2007年 面积：1200平方米

Where the tidal waters of the Atlantic flow into the estuary of the Tagus River, old and new come together in the hotel Jeronimos. Located in Lisbon's historic district and surrounded by museums and monuments, the 1940s building has been given new life by the Lisbon based architects Capinha Lopes & Associates, whose renovation of the hotel liaises with its surroundings. The ornate gables, statues and relief of the nearby 16th century Jeronimos Monastery accentuate the modern renovation. Windows afford a view of the monastery from the hotel's wine bar, where one can taste the rare and much sought, after Bussaco wine that has been produced for generations by the Alex and rede Almeida family, which also owns the hotel. As if to reflect the wines, the bar is designed in a rich savoury palette of red, white and brown, with a dark wood floor and floating lamps.

The welcoming reception desk — a lustrous, dark chocolate cube of wood before a crimson wall — greets guests in the same colour scheme. Each of the 65 rooms and suites has a pale marble bath room generously bathed in natural light. Throughout the hotel, the abundance of natural lighting blends the interior with the Lisbon sky outside.

坐落于里斯本这座古老的城市，被博物馆和名胜古迹环绕，位于大西洋的水流入到塔霍河河口，这就是融合了古老与现代元素的杰尼贸酒店。这座19世纪40年代的古老建筑在凯品哈设计公司的设计下被赋予了新的灵魂，同周围的建筑遥相呼应。装饰华丽的尖顶屋两端的山形墙雕塑以及周围的16世纪的杰尼贸修道院，共同映衬了其翻新后的时尚气息。从酒店酒吧的窗户可以将修道院的景观尽收眼底，在品尝过巴萨克酒之后，这里人们可以享受各色美食，巴萨克酒曾经是亚历克斯家族代代相传的。仿佛是为了衬托葡萄酒，酒吧的主题色调采用艳丽的红色、白色和褐色，深色的木质地板和浮动的吊灯。

酒店接待处的设计是在深红色墙的前面摆放了巧克力形状的、暗色的、光泽感很强的接待台，从色彩上让顾客感觉到一种暖意。65个房间中的任何一个房间和套间都设计有一个浅色的大理石的浴室，可以让您自由地沐浴在自然光之下。整个酒店都沐浴在里斯本天空下的阳光和室内的自然光线之下。

The Levante Parliament

累范特风酒店

Project Name: The Levante Parliament   **Designer:** Michael Stepanek   **Location:** Austria, Vienna
Photographer: Michael Stepanek   **Time:** 2006   **Area:** 4600 sqm

项目名称：累范特风酒店 设计师：迈克尔 斯泰潘内克 项目地点：奥地利，维也纳 摄影：迈克尔 斯泰潘内克
完成时间：2006年 面积：4600平方米

The original building dates back to 1908, and is a perfect example of the Modernist architecture that was initiated by the Vienna School and Bauhaus. Its main characteristics, still evident today, are an emphasis on rationalism, the elimination of ornament and the use of technological advances in materials that allow for flexibility in design. This is the guiding principles by which the multidisciplinary team of architects, designers and artists transformed the building into an innovative space that integrates a gallery and the hotel. The glass design objects by artist Ioan Nemtoi have not only been given 4,600 square metres of prominent exhibition space, Nemtoi was also instrumental in the design of the restaurant and bar areas. The undefined borders between gallery and conventional hotel space have resulted in a challenging fusion of art and contemporary design.

The vision for The Levante's overall design concept is based on an interpretation of the four natural elements – transparency and light represent fire, the natural building materials symbolize earth, the generously sized rooms with 3.3 metres ceiling height stand for air, and the lineal forms and flexibility represent water. A range of core materials including light natural stone, glass, chrome and dark mahogany wood were selected to communicate the relationship to nature and harmonise with the classical modern design of the building.

这个建筑的前身始建于1908年，可谓是现代主义建筑的一个典型的代表之作，为维也纳大学和包豪斯建筑学派所公认。改建后，设计仍然保持原有设计特点，即强调理性，对于装饰的简化以及材料和先进技术的采用，而这些特征使得设计具备了很强的灵活性。在这个理念及原则指导下，不同学科的建筑师、设计师和艺术家将建筑转换成为一个具有创意的空间，将画廊和酒店融为一体。由艺术家亚伊万莫设计的晶状物体不仅占据了整个空间的4600平方米，而且对于餐厅和酒吧的设计也起到了至关重要的作用。画廊与传统的酒店之间模糊的界限是艺术与当代设计的极具挑战性的完美结合。

累范特风酒店的整体设计理念对四种自然元素进行了诠释——通透明亮象征着火，天然的建筑材料象征着土，宽大的空间和3.3米高的天花板意味着空气，线性的窗体和适应性具有水的特性。主要材料的选择包括明亮的天然石头、玻璃、铬、暗色的红木，这些都用以展示着自然的本色，展现着时尚经典建筑与自然的和谐统一。

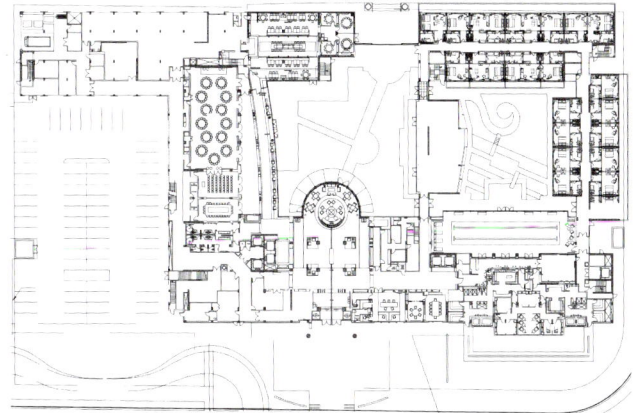

CitizenM Hotel

旅行者酒店

Project Name: CitizenM Hotel   **Designer:** concrete   **Location:** Amsterdam, The Netherlands
Photographer: Ewout Huibers   **Time:** 2008

项目名称：旅行者酒店 设计师：concrete设计事务所 项目地点：荷兰，阿姆斯特丹 摄影：艾沃特·胡博斯 完成时间：2008年

Something exciting is happening in the world of hotels. A new type of traveller is walking through the door. These modern individuals are explorers, culture-seekers, professionals and shoppers. They travel a lot – both long and short haul. They are independent, sharing a respect for the places they visit and are young at heart.

The concept of the hotel is to cut out all hidden costs and remove all unnecessary items, in order to provide its guests a luxury feel for a budget price. The rooms are stacked on a ground floor with a dynamic lobby/living-room space, creating rooms and F&B functions. A red glass box with electronic sliding doors allows entrance to the hotel. The hotel does not have a service desk but a table with six self-check-in terminals. One of citizenM's hosts is present to guide you through the in check-procedure. The public area of the ground floor is divided in several areas to create a home-environment, by designing working areas, dining areas and sitting areas.

现今时代，酒店的住宿客中又增添了新的群体，他们或是探索者、文化追寻者、专业人士，或是购物达人。他们到处旅行，或长期居住，或短暂逗留；他们生活独立，尊重他们居住过的每一个地方。旅行者酒店就是针对这些人群设计的。

为满足客人的要求，理念便是减少隐性成本，摒弃一切不必要的元素，让他们以合理的价格体验奢华感受。穿过带有电动拉门的红色玻璃结构，便可到达酒店大堂。这里没有服务台，取而代之的是设有六个登记通道的大桌子，打造了登机体验。公共空间被分隔成不同的小区域，如办公区、就餐区以及休息区，营造了家一般的感觉。此外，客房紧凑地排列着，亲切而温馨。

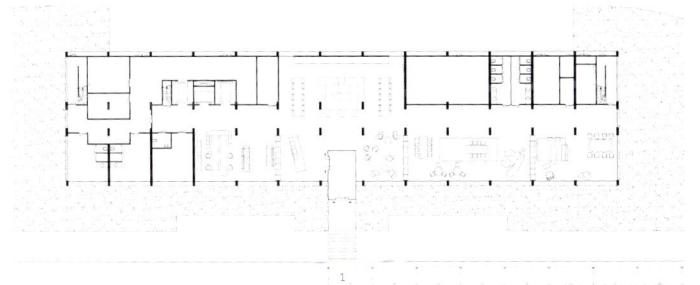

Murmuri Hotel

莫姆里酒店

Project Name: Murmuri Hotel   **Designer:** Kelly Hoppen   **Location:** Barcelona, Spain
Time: 2008

项目名称：莫姆里酒店 设计师：凯莉·霍本 项目地点：西班牙，巴塞罗那 完成时间：2008年

Located in the emblematic Rambla de Catalunya, close to Las Ramblas, La Pedrera and surrounded by be best national and international boutiques, Hotel Murmuri offers those visiting Barcelona an incomparable accommodation choice.

The stone façade and the splendid windows of the building are in harmony with the typical architectonic style of the Eixample quarter. However, the interior has been designed in a wonderfully sober, pleasant and sensual fashion by the acclaimed British interior designer Kelly Hoppen who has managed to take care of every detail and to create a hotel with those personal touches that please even the most demanding and refined guests.

All 53 rooms have been decorated with an exquisite taste, meriting special mention the Suite with views over Rambla Catalunya and the Junior Suite. The combination of dark wood with various hues of beige, cream, grey and brown create a modern, contemporary and sophisticated atmosphere that will "age well".

Treading a thin line between modernity and tradition, Kelly Hoppen has made an intelligent use of neutral colours and rich textures managing to create a style that, though close to minimalism, it is luxurious enough to enchant even the most conservative guests. Her interior design projects have crossed international boundaries; she designed the Beckham's L.A house and even the first class cabins of British Airways.

From the moment a guest enters Hotel Murmuri, (s)he feels that (s)he is in the hands of professionals who not only respond to any of his/her needs, but who are also able to anticipate them. The hotel's team is young, dynamic, international and well trained to offer an excellent service: discreet, approachable and with an attitude and manners that will please even the most demanding guests.

酒店位于加泰罗尼亚大道上，与Las Ramblas大道及米拉之家毗邻，环绕在当地及国际知名精品店之间，旨在为那些游览巴塞罗那的客人们提供一个无与伦比的住宿选择。

酒店石质外观及壮丽的大窗使其完全符合Eixample区的奢华建筑风格。室内装饰恰恰相反，以恬淡、舒适及感性为主要特色。凯莉·霍本注重每一个细节，打造酒店独特的风格，让那些极为挑剔的客人也能得到满足。53间客房的装饰全部以精致的品位亮相，必须提到的是，套房内将加泰罗尼亚大道的美丽景色一览无遗地收纳进来。黑色木头与米色、乳白、淡灰及褐色等多种色调混合，营造了现代典雅的氛围，并且永不过时。

更为重要的是，凯莉·霍本在现代特色和传统韵味中取得平衡，通过运用自然色调及丰富的质感营造了一种独特的风格——与简约风格相近，但又恰到好处地兼顾了奢华。客人从踏入酒店的一刻起，便会充分感觉到其专业的服务，年轻而活力十足的团队让人备感舒适。

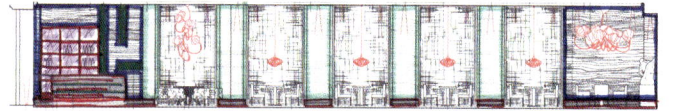

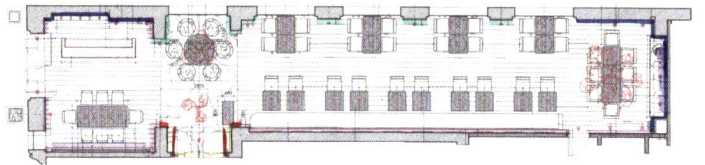

Hotel Sezz

瑟兹宾馆

Project Name: Hotel Sezz   **Designer:** Shahé Kalaidjian, Christophe Pillet   **Location:** Paris, France
Photographer: Christophe Pillet   **Time:** 2005

项目名称：瑟兹宾馆 设计师：沙赫·卡莱吉，克里斯夫托·皮耶 项目地点：法国，巴黎 摄影：克里斯夫托·皮耶
完成时间：2005年

What is it that your first experience at any hotel, no matter how fancy, makes you feel like you're waiting in line at a post office? In order to say no there is Hotel Sezz, which is why there is no check-in desk. When you entry, there is your personal assistant who will be just a phone call away for the length of your stay at the Sezz. You find yourself wishing they would stick around to give you a hand during your stay — making sure things run smoothly, getting you theatre tickets, going shopping with you to show you places not on every tourist map.

It embodies the designer's philosophy about what constitutes a luxury hotel in the 21 century. It's not about teak furniture, cashmere throws on the bed or hand-blown Murano glass flower vases. As it happens, the vases in the Sezz are hand-blown Murano glass. But as far as the designer is concerned, fine materials are just the opening ante to sit at the table. The smallest room is 200 square metres, and the larger ones are 400 square metres.

Running a luxury hotel, the designer promotes ease, comfort and unforced elegance and directed that most beds in the Hotel Sezz stand freely in the middle of the room, the easier to stroll around them. But really, the look of luxury is about a lot more than buying nice lamps and chairs. The trick is pulling everything together in a seamless whole.

无论多高档的宾馆，您进去的第一印象是什么？是不是感觉登记时就像在邮局排队一样？为了对这种现象说"不"，瑟兹宾馆就没有设置接待台。当你进入宾馆时，就会有一个你自己的私人助理，他就像你在宾馆内的一部移动电话，随叫随到。你就会发现自己希望他们能停留在你身边帮助你——确保各项事情都顺利，帮你买电影票，陪你逛街，并告诉你旅游地图上没标注的地点。

这个宾馆表现了设计者关于设计21世纪奢华宾馆的设计哲学，这并不关柚木家具，床上的山羊绒毯子或是人工吹制的慕兰奴玻璃花瓶的事。虽然在宾馆里摆放着慕兰奴玻璃花瓶，但正如设计者所考虑的,优秀的材料正如桌边的室外天线——唾手可得。宾馆内最小的房间有200平方米,大些的有400平方米。

经营一家奢华酒店,设计者提倡安逸、舒适和自然的优雅感,并指出,酒店内大多数的床都要布置在房间中间,这个可以更容易的绕着它们散步。但是说真的,看起来奢华并只是买些漂亮的灯和椅子,秘密在把所有的东西做为一个完美无缺的整体摆放一起。

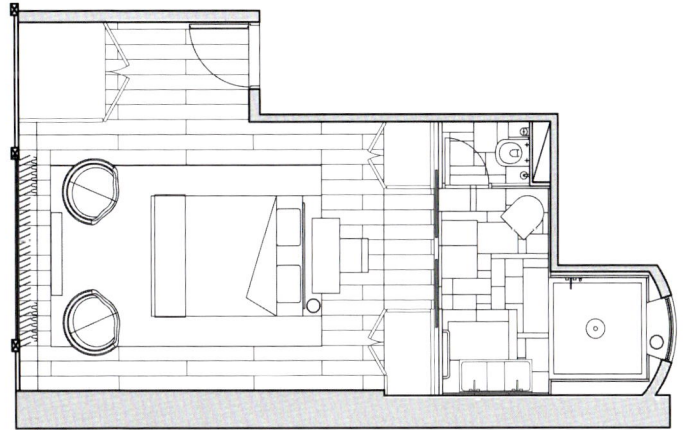

The George Hotel

乔治酒店

Project Name: The George Hotel   **Designer:** synergy hamburg   **Location:** hamburg, germany
Photographer: synergy hamburg   **Time:** 2008

项目名称：乔治酒店 设计师：汉堡协力设计室 项目地点：德国，汉堡 摄影：汉堡协力设计室 完成时间：2008年

The George's unpretentious façade melds into the district's urban scene. Once inside, guests are warmly greeted by friendly staff and a welcoming atmosphere. "The design is meant to be fun", says Managing Director Kai Hollmann. "Stylistic contradictions here and there are absolutely intentional." The primary design scheme is held in black and white, yet the contrast is softened with warm dark woods and spots of colour. These heavier elements are skillfully balanced with gauzy fabrics, glass and an intelligent lighting design. Craftsmanship plays an important role in the interior design and handmade fabrics, carpets and unique pieces were selected by Hollmann during his travels.

A tour through the hotel is like a stroll through London: inspiring and visually varied. Exotic influences dating back to colonial times make a striking contrast to modern classics. Indian saris are colourful highlights as bedspreads while an oriental Marrakech style reigns in the spa. In the Library, modern interpretations of Chesterfield furniture, cosy niches as well as lavish carpets give the space a clubroom feel. It's the perfect spot to relax or read…and of course indulge in The George's own rendition of Afternoon Tea.

The hotel developer Kai Hollmann invests much dedication and commitment to his projects including sourcing unusual design pieces for his hotels through fine antiques dealers and specialist furniture boutiques.

酒店的外观朴实无华，毫不起眼。进入酒店，客人就会受到热情的接待，身处温暖的氛围之中。"饭店的设计目的就是要有趣"，总经理凯·霍尔曼说，"很多矛盾都是我们特意设计出来的"。设计基调以黑色和白色为主，采用黑色木纹和斑点图案缓和黑白之间的反差。设计师巧妙地运用轻纱、玻璃和智能照明来搭配浓重的色彩。手工艺品在室内设计中起重要作用，这里的手工织物、地毯和别致的摆设都是霍尔曼在旅游时淘来的。

穿梭在酒店之中尤如漫步伦敦街头，给人以不一样的感受。异国情调的装饰方法可以追溯到殖民时期，它与现代的经典风格截然不同。卧室的床单由色彩鲜艳的印度纱丽制成，亮丽夺目。水疗馆是东方马拉喀什风格。图书馆则用现代手法来诠释，切斯特菲尔德的家具、温暖的壁龛和豪华的地毯把这里装点得如同俱乐部的聚会室一般。图书馆是放松与阅读的绝佳场所，再来一杯下午茶，身心都陶醉其中。

开发商凯·霍尔曼为酒店投注了大量心血，他与大古董商和家具精品店合作，使饭店内的家具和摆设都很与众不同。

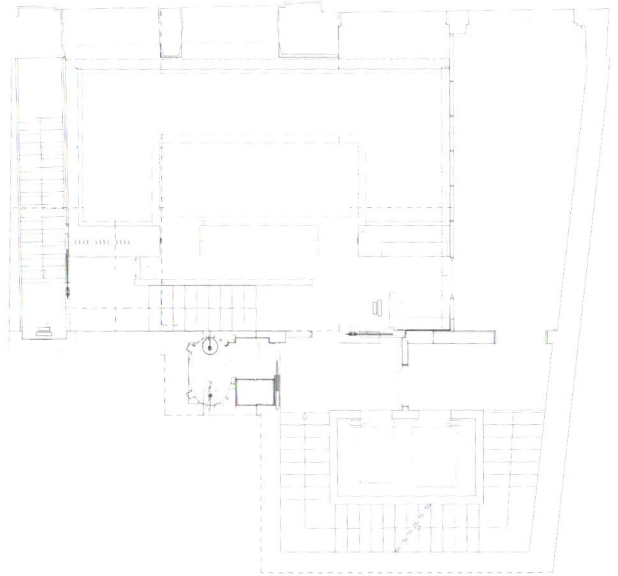

Villa Florence Hotel

佛罗伦萨别墅饭店

Project Name: Villa Florence Hotel   **Designer:** sfa design   **Location:** San Francisco, California
Photographer: Ken Hayden   **Time:**2008

项目名称：佛罗伦萨别墅饭店 设计师：什方设计公司 项目地点：旧金山，加利福尼亚 摄影：肯海登 完成时间：2008年

Inspired by private homes in Florence and Tuscany, sfa sought to create a chic, modern design that would honor the most elegant, alluring elements of a traditional Italian villa. Mixing modern furniture with classic architecture, sfa design created a relaxing home away from home for anyone seeking a more personal escape in an expansive urban setting. Crossing San Francisco's swarming Powell Street, hotel-goers now meet an entrance of stone mosaic flooring and a rustic rock wall covered in ivy vines and an antique fountain. Stepping onto the Dijon-mustard, gold and cream terrazzo flooring of the foyer and greeted by an old painted mural of Florence, the ambiance is one of an expensive, yet intimate Italian home. Reinterpreted modern Italian art and contemporary velvet sofas adorned with colorful pillows invite guests into the warm living room of a private villa. Whimsical rugs and eclectic accessories from various antique stores and odd shops further this sense of an isolated Italian dwelling, always with a hip nod to the chic influences of modern boutique design. Clean-lined reception pods feature custom living glass panels lit from above to illuminate encased silk, vintage Italian scarves. The guestrooms are equally joyful and surprising in a palette of terracotta and celadon. Photographs of the familial scenes, modern art photography and landscapes give each room the feeling of home, while fun vintage images like that of Sophia Loren liven up the luxuriously modern bathroom.

设计师受佛罗伦萨和托斯卡纳私人住宅的启发，运用时尚现代的设计手法，创造了一幢高雅迷人，且具有传统意大利风格的别墅。设计师把现代家具与古典建筑相结合，创造了一个舒适的家园，一块闹市中的净土。穿过旧金山繁华的鲍威尔街，别墅饭店即映入眼帘。入口的地面铺着马赛克，质朴的石墙上爬满了常春藤，门口还有一个古老的喷泉。进入大堂，踏上绿色、金色和米色水磨石拼成的地板，迎接您的是一幅古老的佛罗伦萨壁画，这里有意大利式奢华而温馨的家庭气氛。现代的意大利丝绒沙发上放着五颜六色的靠垫，邀请来宾到温暖的客厅里坐坐。从古董店和特色店里淘来的别致的地毯和摆设使别墅成了一幢正宗的意式住宅，并且全部采用现代精品设计。接待处十分简洁，定制的玻璃天花板下挂着古香古色的意大利丝质缎带，在灯光下闪闪发亮。客房以土红色和青绿色为主，看上去令人心旷神怡。每间客房的墙上都挂着家族照片和现代艺术摄影作品，窗外都有美丽的风景，给人以家的感觉。现代的豪华浴室中挂着老旧的漫画，大概是索菲亚·罗兰时期的作品，为浴室增色不少。

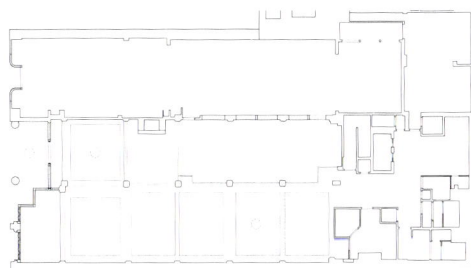

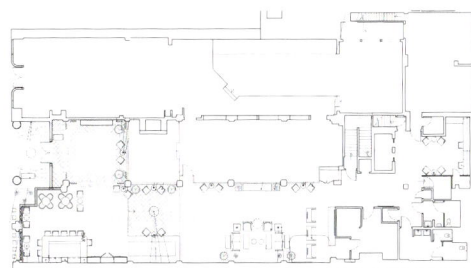

JW Marriott Hotel Hong Kong

香港JW万豪酒店

Project Name: JW Marriott Hotel Hong Kong   **Designer:** JW Marriott Hotel Hong Kong
Location: Hongkong, China   **Time:** 2007

项目名称：香港JW万豪酒店 设计师：香港JW万豪酒店 项目地点：中国，香港 完成时间：2007年

Located near Hong Kong's International Airport and AsiaWorld Expo, the property's location gives easy access to the CityGate Outlets, Ngong Ping 360 cable car and the Tian Tan Buddha. Hong Kong's Central district is a 25-minute express train ride away.

The 658-room hotel is the only property in Hong Kong to offer guests golf packages with access to the adjacent SkyCity Nine Eagles golf course, and is the first to offer guests the territory's only Marriott spa, Quan Spa.

The façade flows with luxury to correspond to the flourishing context of the city. At the same time, the "luxury" concept continues in the interior — every element being meticulously selected and every detail being carefully handled. Comfort is another concept that the design concerns to ensure the guests a relaxing atmosphere. The bar in the hotel boasts sumptuousness and gorgeousness. The ceiling is dotted with "shining stars" with the beautiful chandeliers hang below, showing the "luxury" concept to the root; the seat made in unique shape also serves as decorative element. What more attracts eye is of course the lighting — shinning on the wine bottles.

The guestroom highlights comfort. The large French windows are specially designed from which natural light come in to render the whole room emerged in sunshine. Moreover, from here, one can fully enjoy the beautiful landscape outside. Moreover, the combination of red and white colours further enhances the concept.

JW万豪酒店与香港国际机场及亚洲国际博览馆毗邻，地处通往东荟城名店、昂平360缆车及天坛大佛的必经之路，距中环仅为25分钟车程。

酒店共包括658间客房，是香港地区唯一一家为客人提供高尔夫门票及装配以及水疗服务的酒店。

酒店外形宏伟豪华，让人感受都市的繁华。内部设计也秉承了万豪酒店一贯的奢华风格。设计师精心挑选了室内的每一个装饰材料，精心设计每一个细节，装饰了室内环境的同时也给顾客带来舒适的体验。

酒吧的设计热烈而不乏华美，屋顶的设计犹如繁星点点，再搭配精致的吊灯，彰显贵族气息，造型别致的座椅同时起到了装饰的作用。酒吧中的灯光设计是一大亮点，各色灯光照射着酒柜中的酒，闪闪发光，非常引人注目。

客房设计很舒适，设计师选择了极富贵族气息的红色系与白色搭配，每一件家具的选择都体现了酒店不凡的品位。客房都有开敞的落地窗，让客房沐浴在阳光里，也可以欣赏城市的美景。

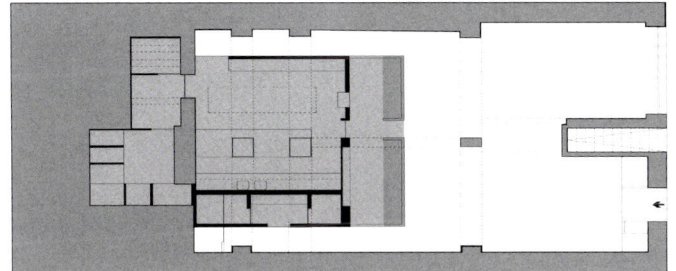

Hotel Pirámides Narvarte

那瓦特金字塔酒店

Project Name: Hotel Pirámides Narvarte   **Designer:** DIN interiorismo, Aurelio Vázquez Durán
Location: México City, México   **Photographer:** Jaime Jacott   **Time:** 2007

项目名称：那瓦特金字塔酒店 设计师：DIN室内设计 项目地点：墨西哥，墨西哥城 摄影：杰米·加考特 完成时间：2007年

The main design concept was to create an oasis within Mexico City. The interior design is a retreat in where warmth and harmony are the leading characters of the ambiance. The hotel has 83 rooms and 10 different layouts were designed to have an interesting variety of regular rooms and suites. All the furniture is placed with a 30° inclination to avoid the monotony and interact with the different spaces that each room and suite has. Custom design art-object was designed and produced specially for each room. Wood frames with high resolution digital printed photographs of abstract parts of the human body hang from the walls and match with the tables that have the same kind of images.

Different levels generated by stairs and platform add dynamic to the rooms and empathise the different areas. The bed is always located in the centre of the rooms leaving the rest of the furniture to create the other spaces around it. Natural fiber hammocks, dinning room and a Jacuzzi with deckchairs complement the relaxing areas.

The lighting design was done with theatrical effects in mind. The combination of dimmers and different light temperatures create the atmosphere necessary for each activity while in the room. All the furniture and surfaces have a special UV coating for easy cleaning and maintenance. Bed covers and curtains were selected in a color scheme that includes: brick red, pistachio, cream and almond coffee. This colour palette is complemented with the natural materials and finishes of all the areas such as wood, granite and marble.

酒店主要设计理念是打造城市中的绿洲，在内部装饰上为客人营造一个温暖而和谐的氛围。酒店共有83间客房，设计师为此设计了10种不同的格局，实现多样性要求。所有的家具都呈30度角倾斜，打破了单调感；每间客房内都悬挂着专门定制的艺术作品；人体素描图画镶嵌在木质框架内与带有同样图案的桌子交相呼应。

客房内，台阶和平台的设置增添了动感，同时将不同的区域区分开来。床被摆放在中央，其余家具围绕其展开，形成了不同的空间。天然纤维吊床、餐厅以及带有帆布躺椅的JACUZZI浴缸使得休闲区更加完善。

独特的灯光设计营造了舞台般的氛围，调光器以及光温调控设备的运用则创造了可变换的环境，格外舒适。所有的家具及表层结构全部采用特制的UV覆层，便于清洗和维护。床罩及窗帘的选择注重色彩搭配，木材、花岗岩及大理石材质的应用与空间色彩相得益彰。

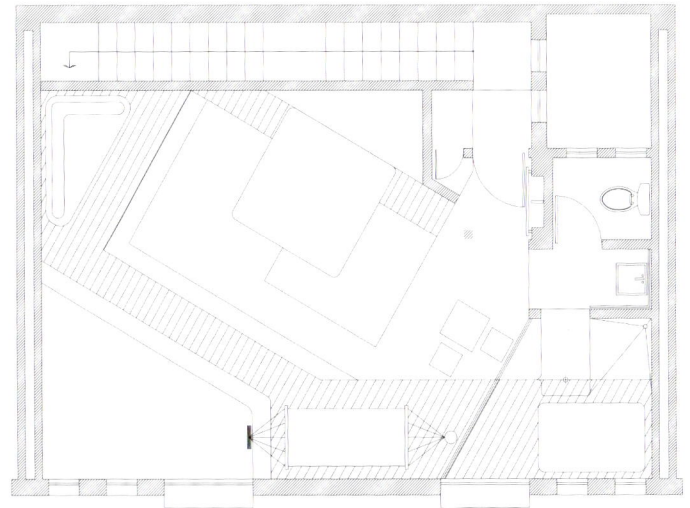

Mövenpick Airport Hotel

慕温匹克机场酒店

Project Name: Mövenpick Airport Hotel **Designer:** Matteo Thun & Partners
Location: Stuttgart, Germany **Photographer:** Matteo Thun & Partners **Time:** 2007

项目名称：慕温匹克机场酒店 设计师：马泰奥·图恩合作事务所 项目地点：德国，斯图加特 摄影：马泰奥·图恩合作事务所
完成时间：2007年

The way Mövenpick Hotel has been dressed up in grey pinstripe with touches of Prada Green provides that something extra perceptually speaking. Added symbolic value comes from the way the place is felt: airports are nodes on an international network, so the genius loci is a sort of national banner, not a local touch. The rooms in particular, provide places where guests retire, work on this assumption: it is the nation, not the city, which provides the most emblematic icons in a collage of images.

慕温匹克酒店灰色细条纹的设计带有一丝绿色普拉达的风情，显示出这里在感知上的超凡之处。另外，酒店还具备象征意义，这从酒店的所在地上能够感受出来：机场是国际网中的节点，所以在一定程度上代表了国家的形象，而不只是当地。尤其是酒店客房，是客人休息、工作的私人场所，从酒店所在地的角度上考虑，也是代表了整个国家的形象，而不只是这座城市。

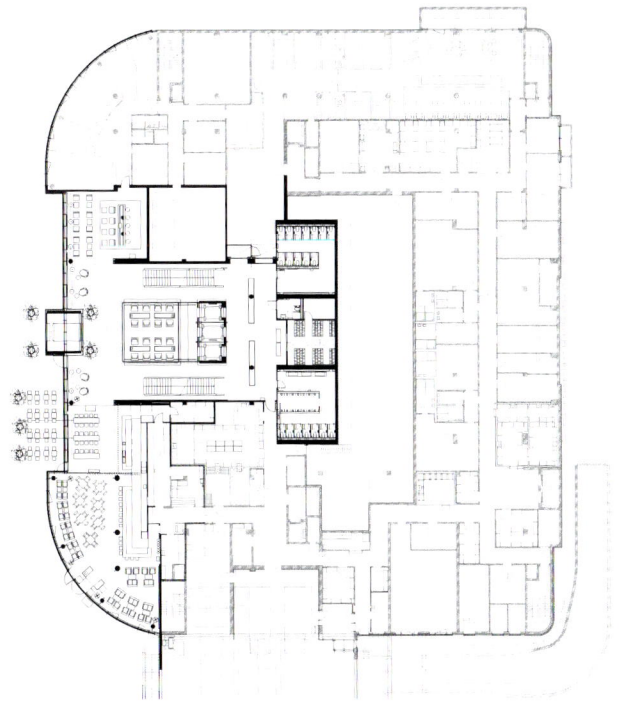

The Library

图书馆大酒店

Project Name: The Library **Designer:** Tirawan Songsawat **Location:** Ko Samui, Thailand
Photographer: Tirawan Songsawat **Time:** 2008

项目名称：图书馆大酒店 设计师：蒂拉万·颂萨瓦特 项目地点：泰国，苏梅岛 摄影：蒂拉万·颂萨瓦特
完成时间：2008年

Distractions are kept to a minimum at this graceful hotel complex set in 6,400 square metres of lush Thai beachfront in Koh Samui. Group designer Tirawan Songsawat has created a minimalist structure, preserving a heritage property at the water's edge while intruding as little as possible on its ecology and aesthetics. The hotel's 26 suite-studio cabins are scattered discreetly around the ground. These consist of a ground-floor suite space and a separate studio upstairs that offers fabulous views of both ocean and old-growth trees that have been spared the developer's bulldozer.

Guests may lose (or find) themselves among vegetation interspersed with artwork and statuary, contrasting a bold colour scheme in which white, red, black and grey predominate: the Library's exterior is white, the swimming pool red, the restaurant grey. Interiors follow the same ultra minimalist palette and feature low-slung, rectilinear furnishings that invite visitors to stretch out and listen to the lapping of the waves – or of course dive into the books in the exemplary namesake library. Plasma-screen televisions and broadband connections complement golden Buddhas and wooden shutters in a perfect balance of nature, comfort and art.

Set on the beachfront area, the Page Restaurant and Bar provides an air-conditioned area and an open-air space overlooking the sea and the vivid red swimming pool. You can also relax on the wooden deck around the pool and restaurant area furnished with soft cushions and mattresses serving as sun loungers.

这座高雅的酒店建筑群占地6400平方米，坐落在泰国苏梅岛的海滨，远离纷扰。设计师Tirawan Songsawat打造了一座极简主义建筑，保护了这座水滨文化遗产，同时对其生态、美学方面尽量造成较小影响。酒店包含26个"套房加工作室"，精心布局，其中包含一楼的套房空间和楼上的工作室。工作室拥有望向大海和古树的绝佳视野——幸亏古树逃过了开发商的推土机。

酒店客人将置身于一片绿色植被当中，里面穿插点缀着艺术品和雕塑，跟大胆的色调——白色、红色、黑色和灰色——形成鲜明对比。酒店的外观是白色的，游泳池是红色，而餐厅则是灰色。室内延续了同样的极简主义风格色调，以低矮的流线型家具摆设为特色，让参观者不禁伸个懒腰放松一下，一边听着水波拍击的声音，或者当然也可以一头扎进图书馆中。液晶电视和宽带配合金色的佛像和木质百叶窗，在自然、舒适和艺术之间达到了理想的平衡。

"书页"餐厅和酒吧也坐落在这海滨之地，是酒店里的一个凉爽、开放的空间，俯瞰大海和鲜艳夺目的红色游泳池。你也可以在泳池和餐厅周围的木板上放松一下，这里配有柔软的靠垫和床垫，可以用作躺椅。

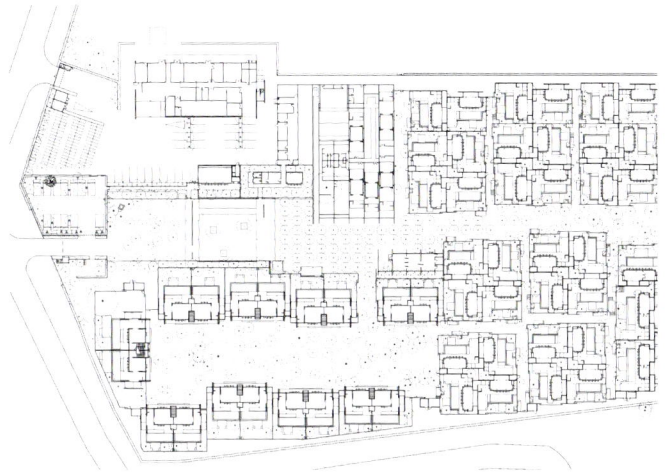

X2 Koh Samui

苏梅岛X2酒店

Project Name: X2 Koh Samui   **Designer:** Be Gray Limited   **Location:** Koh Samui, Thailand
Photographer: Be Gray Limited   **Time:** 2008

项目名称：苏梅岛X2酒店 设计师：毕格雷有限公司 项目地点：泰国，苏梅岛 摄影：毕格雷有限公司 完成时间：2008年

The X2 Koh Samui resort, pronounced "crossto", is intended for a cosmopolitan crowd eager to avoid the strict formality of traditional highclass hotels. Eager, that is, to "cross to" a whole new dimension of laidback luxury.

Beautifully located on Hua Thannon beach on the Southeastern coast of Koh Samui island, the newly-built resort offers twenty-seven villas resting on a largely undeveloped beach surrounded by old tree growth. The buildings' sleek, horizontally oriented design seems to meld into nature while its interiors in calm neutral tones stand in sharp contrast to the Gulf of Thailand's brilliant turquoise waters. Most of X2's villas have their own swimming pools and terrace gardens, set against a backdrop of swaying unspoilt palms. There is plenty to do and even excellent fusion food on offer all day at the resort's 4K (pronounced "fork") restaurant, but the resort's main activity is relaxation. For this reason, the huge outdoor massage pavilion probably best captures X2 Koh Samui's true essence. With just two beds on which to be pampered by expert hands, the coolly modernist building has an entire acre of land and thirty metres of beachfront to itself. This is a place to really feel that the most desirable luxuries of today are infinite time, endless space and sweet silence.

At 4K Restaurant, international executive chefs offer a wide range of fusion European and Thai dishes. The 4K restaurant and bar at X2 provide all day dining for guests between 7:00 am and 11:00 pm.

苏梅岛X2度假酒店专为大都市人群设计，他们渴望避免传统的高级酒店中严格、拘泥的形式。他们渴望的，是一种全新的维度，一种从容的奢华。

这家新落成的酒店坐落在苏梅岛东南海岸美丽的海滩上，包含27座别墅，这是一大片尚未开发的土地，周围是生长茂盛的古树。酒店的设计以横向为主，造型优美，看上去似乎跟周围的自然景观融为一体，而室内采用的是淡雅的色彩，跟泰国湾明媚的青绿色海水形成鲜明对比。别墅大多拥有自己的独立泳池和房侧花园，在一片未受破坏的迎风摇曳的棕榈树中间。这里有很多好玩的，而且4K餐厅全天都有美食供应，但是酒店最主要的活动仍是休闲。为此，宽敞的户外按摩亭可以说是酒店最具特色的地方。亭内只有两张按摩床，客人可以躺在上面，享受专业手法的按摩。按摩亭采用冷峻的现代主义风格，占地一英亩，并且享有长达30米的海滩。这个地方能让你真正感受到今天的奢华意味着无限的时间、空间和甜美的静谧。

4K餐厅里，国际名厨将为您准备各种欧洲美食和泰国菜。X2酒店里的4K餐厅和酒吧全天候为客人供应美食，从早7点到晚11点。

Thurles Arts Centre and Library

瑟勒斯艺术中心及图书馆

Project Name: Thurles Arts Centre and Library **Designer:** McCullough Mulvin Architects
Location: Thurles, Ireland **Photographer:** Christian Richters **Time:** 2006

项目名称：瑟勒斯艺术中心及图书馆 设计师：罗及莫尔文建筑事务所 项目地点：爱尔兰，瑟勒斯 摄影：克里斯蒂安·瑞契
完成时间：2006年

In plan the building is similarly cranked, each zone mapped into trapezoidal volumes which master the bend of the river. In the library, a long thin space with its volume pressed to the river, a deep cut in the ceiling plane right through the research floor brings light and air to the centre of the plan at reading spaces and the issue desk. The exhibition space has a similar slice through the roof plane orthogonal to the first to conduct daylight through a huge roof light sitting across the upper terrace, giving unexpected views of the work on exhibition below. An introverted, reflective space, its walls splay out towards the riverfront, taking up the geometry of the site. Shielded behind the monolithic concrete entrance wall, the space can be glimpsed through a porthole when arriving, or alternatively closed off for hanging. The theatre foyer is similarly a compressed volume – vertical this time – caught between auditorium and boardwalk. Piercing the heavy concrete wall of the theatre, the control room is suspended over the café/bar. Large glass doors slide back to open the café and foyer to the boardwalk, and from the upper foyer the audience expands out on to the upper terrace overlooking the town.

Throughout the building colour is used to code and focus – red for information, orange for vertical circulation, white and black for concentration and relaxation. Seen from across the river these coloured zones resolve themselves into large scale gestures of connection in the case of tubes of orange stairs rising diagonally behind glass, intense dots of red marking out information areas, calm white zones of research and introspection, spots of black indicating more expansive night uses.

按照设计规划，这座建筑内的每个空间都成梯形，展望河套。图书馆内，狭长的区域紧挨着河流，天花板上深深的开口直通上面研究区的地板，让阳光和空气能到达下面中心的阅览区和讨论台。展示区的屋顶也有类似的开口，与图书馆屋顶的开口成直角，使人坐在楼上的露台上就能望见下面的展示作品。这里是一个适宜深思的地方，墙体像河边张开，顺应建筑地的几何形状。大门口的单体水泥墙后面隐藏着的空间，来访者可以透过舷窗一窥，有时也挂上帘子挡上。剧场的门厅是一个压缩空间——这次是垂直压缩——在礼堂和木板小道中间。水泥墙体的剧场上面，控制室悬在咖啡馆之上。巨大的玻璃拉门连接咖啡馆、门厅和木板小道，而且从上面的门厅，观众可以上到露台上俯瞰城镇。

整个建筑对颜色的运用既达到了统一，又突出了焦点——红色用作信息，橘黄用作垂直通道，白和黑分别用作学习和娱乐。从河上看去，这些颜色鲜明的区域交互掩映：橘黄的楼梯在玻璃墙后垂直穿过；一点点醒目的红色标示出信息区；沉稳的白色代表了研究和思考；黑色表达的则是"夜晚"，这里更广泛的使用。

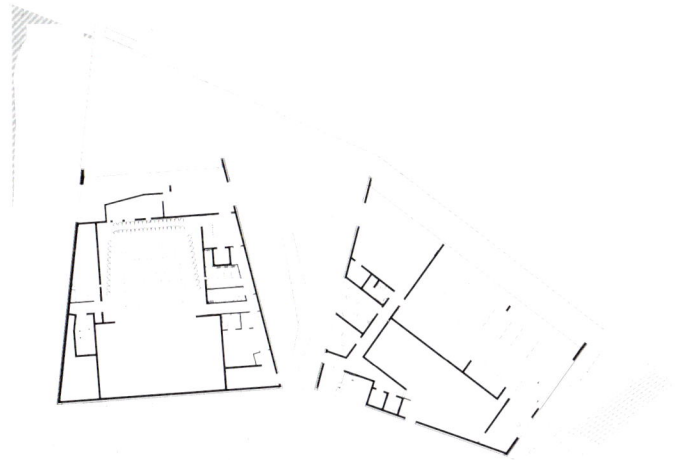

A.E. Smith High School Library

阿尔弗德·E·史密斯高中图书馆

Project Name: A.E. Smith High School Library **Designer:** Atelier Pagnamenta Torriani
Location: New York, USA **Photographer:** Frank Oudeman **Time:** 2008

项目名称：阿尔弗德·E·史密斯高中图书馆 设计师：帕格纳曼塔—托里亚尼工作室 项目地点：美国，纽约
摄影：弗兰克·欧德曼 完成时间：2008年

The Alfred E. Smith High School building was built in the 1930s in Bronx as a vocational high school, providing its students not only with a high school diploma, but also with vocational skills such as building construction and auto repair. In fact, the school has gigantic car workshops. The project is the total reconstruction of an existing library space.

Public Schools in New York City are generally very large standardized brick buildings and have an average of 1100 students. The classrooms are situated along a double loaded corridor and the library space is part of the sequence.

The goal is to transform each new library into the school's living room.

A space that is enjoyed by the students, as well as the teachers and parents: kind of the heart of each school.(In fact, when the library had to be closed for the photographic shoot, the students, the teachers and even some parents asked to use the space.)

The designers incorporated into the existing room a floating element that modifies the library space perception and highlights the arts and crafts aspects of the school. The soffit is designed as a geometric progression: a double helix repeating modules of 12 triangular facets and defines the dynamics of the reading areas. The library-flooring pattern mimics the triangular progression.

Low open shelving separates the classroom, the lounge, and the computer area, leaving clear sightlines for the librarian. All furniture is modular, light and stackable; therefore, the space can morph from a classroom to an informal crescent layout to a formal meeting area for the teachers and parents.

Materials and finishes used in the project are environmentally friendly, recyclable and contain low VOC. With the aim of spending energy where it matters, Atelier Pagnamenta Torriani has put a good deal of time into public schools in order to provide quality for those who can benefit the most.

阿尔弗德·E·史密斯高中位于布朗克斯区，除开设高中课程之外还向学生传授建筑及汽车修理等职业技能（拥有较大的汽车修理车间）。学校大楼始建于20世纪30年代，这一工程的主要目标即为翻新现有图书馆。

纽约公立学校在建筑上采用统一标准——大规模的砖石大厦，平均可容纳1100名学生，教室沿着走廊两侧设置，图书馆也位列其中。图书馆的设计要求是打造一个如同起居室般的空间，为学生、老师和家长营造一个舒适的氛围。

设计师从这一点出发，在原有空间内增添了一个悬浮的拱腹结构，增强空间感同时突显学校的艺术特色。拱腹结构由一系列连续的几何形状构成——有12个长方形模块反复叠加构成双链螺旋造型，为阅读区带来活力。地面图案与拱腹相互呼应。低矮的书架将教室、大厅、计算机区分离开来，让图书管理员的视线能够遍及每个角落。所有的家具都由模块搭造而成，轻巧而便利，可以方便实现空间功能的转换。

材料选择上强调环保理念，可循环利用材料及低挥发性有机材料被大量运用进来。

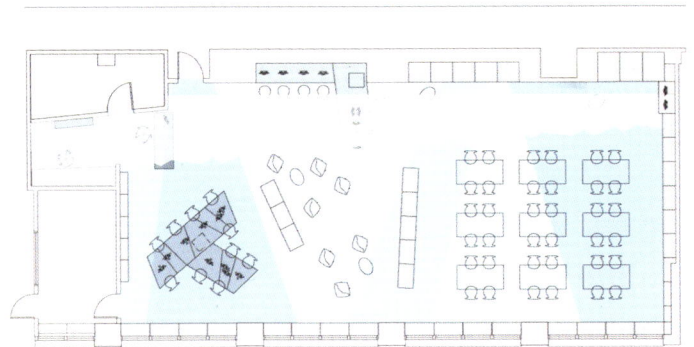

Wagner Middle School Library

瓦格纳中学图书馆

Project Name: Wagner Middle School Library **Designer:** Atelier Pagnamenta Torriani
Location: New York, USA **Photographer:** Frank Oudeman **Time:** 2007

项目名称：瓦格纳中学图书馆 设计师：帕格纳曼塔—托里亚尼工作室 项目地点：美国，纽约 摄影：弗兰克·欧德曼
完成时间：2007年

The Wagner Middle School building was built in 1960 on the Upper East Side of Manhattan for junior high school students. The library space faces the playground and offers beautiful views of the townhouses along the street. The project is the reconstruction and enlargement of a library space.

The librarian requested shelving, a classroom, computer areas, a media centre, lounge area and a librarian desk. The designers manipulated the space by creating a series of waves, inspired by Hokusai's painting: Great Wave at Kanagawa. The soffit, the shelving and the floor pattern mimic the waves generating a spatial dynamic between the circulation and the reading areas.

The library is generally subdivided into four main functional areas: the greeting area with the librarian's desk, the computer area, the classroom and the lounge (chill-out) area. Low freestanding shelving units in the main library space define the various areas.

Lighting can be adjusted in each zone to create spatial effects, so that the users can emphasise one area or create a general subdued atmosphere when viewing a film. The graphics are designed with the principals and librarians. In the case of the vocational school, the principal desired a strong three dimensional soffit, rather than graphics.

瓦格纳中学大楼选址在纽约上东城区曼哈顿，始建于1960年。图书馆朝向大操场，将沿街一侧的联排住宅景致一览无余地收纳进来。这一工程主要任务即为图书馆的重建及扩建。

管理员要求使用橱架结构将教室、计算机区、媒体中心、休息室及管理员工作区分隔开来。设计便从此展开——他们在空间内打造了一系列的波浪造型（源于葛饰北斋的名画——《神奈川上的大海浪》的影响）。之后，拱腹、橱架及地面图案全部模仿波浪造型打造，增添了通道区及阅读区之间的动态感。

图书馆中心区域内采用低矮的、可自由移动橱架连结，同时界定了不同的空间——接待区、计算机区、教室及休息室。其中，每个区域内的光线可自动调节，便于打造不同的空间效果。比如说，在看电影的时候就可通过光线变换强调某一区域，进而营造独特的氛围。平面图案全部由校长和图书管理员亲自设计。

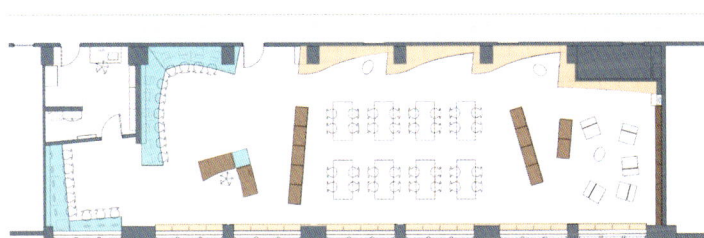

Ps 11r Primary School Library

Ps 11r小学图书馆

Project Name: Ps 11r Primary School Library **Designer:** Atelier Pagnamenta Torriani
Location: New York, USA **Photographer:** Atelier Pagnamenta Torriani; W. Tan, J. Son **Time:** 2007

项目名称：Ps 11r小学图书馆 设计师：帕格纳曼塔—托里亚尼工作室 项目地点：美国，纽约 摄影：帕格纳曼塔—托里亚尼工作室 完成时间：2007年

PS 11r is a primary school, built in 1919, in a middle class residential neighborhood in Staten Island, New York. It has beautiful large wood windows that allow natural light to penetrate deep into the classrooms. The library space is situated just above the auditorium roof and has unobstructed views to the green hills beyond.

A primary school library has to provide not only books and a classroom, but include space for the small kids to play. Therefore, the designers incorporated in the existing room an activity ribbon that separates the main shelving from the classroom. The activity ribbon provides a continuous lower area for the kids to sit and play, as well an area for the librarian. Above the ribbon, a deep soffit hides the air conditioning units. At both ends of the ribbon, two storage closets with a special whiteboard finish, provide a useful writing surface for fun activities. The open wood benches placed along the ribbon offer seating for the kids during the story hour.

Along the soffit, facing the classroom, the sentences from classic children books are playfully displayed while the names of the authors can be discovered at the back. The integrated shelving wraps around the whole room and has a spiral display section that from the lower level winds up to the top. The small classroom space is flexible and the light furniture can be re-arranged easily for parent-teachers meetings.

Ps 11r小学位于纽约史坦顿岛一个中产阶级住宅区内，建于1919年。光线穿过高大的木窗照射到教室的每一个角落，图书馆位于礼堂上方，将周围青山的美丽景致尽收眼底。

小学图书馆有其自身独特的需求，除为学生提供书籍及教室之外，还应设置供低年级学生玩耍的空间。设计师从这一点出发，在空间内划分出一个活动区，将图书区同教室分隔开来。活动区内，低矮区域供孩子们坐下来休息或玩耍，另外的区域则专为管理员使用。活动区上方，极深的拱腹结构将空调等"隐藏"在内；两侧，储存箱表面可用作白色书写板，供孩子们涂鸦。沿着活动区摆放的木椅为孩子们提供了听故事的场所。

另外，沿着拱腹并朝向教室的一侧展示着儿童书籍中的经典词句，背面则书写着作者的名字。整体橱架将整个空间"包裹"起来，并形成了螺旋状的展示区，别具特色。小教室空间布局突显灵活性，通过家具的组装及重新摆放便可为家长会提供场所。

The Danish Jewish Museum

丹麦犹太人博物馆

Project Name: The Danish Jewish Museum **Designer:** Tomrerfirma Gert Fort A/S
Location: Copenhagen, Denmark **Photographer:** Tomrerfirma Gert Fort A/S **Time:** 2007

项目名称：丹麦犹太人博物馆 设计师：汤姆莱费玛·格特·福特，A/S公司 项目地点：丹麦，哥本哈根
摄影：汤姆莱费玛·格特·福特，A/S公司 完成时间：2007年

The architecture of the interior entrance space is meant to communicate the true importance of the museum. The visitor is drawn into the internal courtyard entrance, marked on both the horizontal and vertical dimensions. The horizontal space, or ground level of the entrance, is configured by an ensemble of conversation spaces developed into intimate meeting points for visitors and a space for an outdoor cafe in the summer months. The vertical walls are then marked by a projection of the Mitzvah configuration whose trace can be followed into the depths of the exhibition.

The organising principle of The Danish Jewish Museum is the concept of Mitzvah and its deep ethical meaning as a commandment, a resolve, and as a fundamental good deed. The museum takes the tradition of writing, reading and memory as the overall matrix of organising the exhibition space. In doing this, it is Mitzvah, on both emblematic and architectural levels, that guides a dialogue between the ancient vaulted space of the Royal Boat House and the walls of the Royal Library in relation to the experience of the new museum. To further emphasise this idea the entire exhibition space is illuminated by a luminous stained glass window that is a microcosm of Mitzvah, transforming light across the day.

建筑的入口显示着博物馆的重要性。入口处的庭院十分宽敞。这里不再是仅供来宾交流的空间，它的私密性有所增加，还设有一个夏季开放的户外咖啡馆。建筑的墙壁体现了犹太教"行善"的宗旨，这一宗旨贯穿整个展览的始终。

丹麦犹太人博物馆是在"行善"的基础上建立的，这一词的深层涵义是诫律、坚毅和做善事。博物馆将传统的读、写、记忆融入整个展览空间之中。在其象征意义和建筑水平上都做到了"行善"，将皇家游艇展厅那古老的拱形屋顶和皇家图书馆的古老砖墙与崭新的博物馆完美地结合在一起。阳光透过彩色玻璃窗，把整个博物馆笼罩其中，进一步强调了这一宗旨。这正是"行善"的缩影，即为世间带来光明。

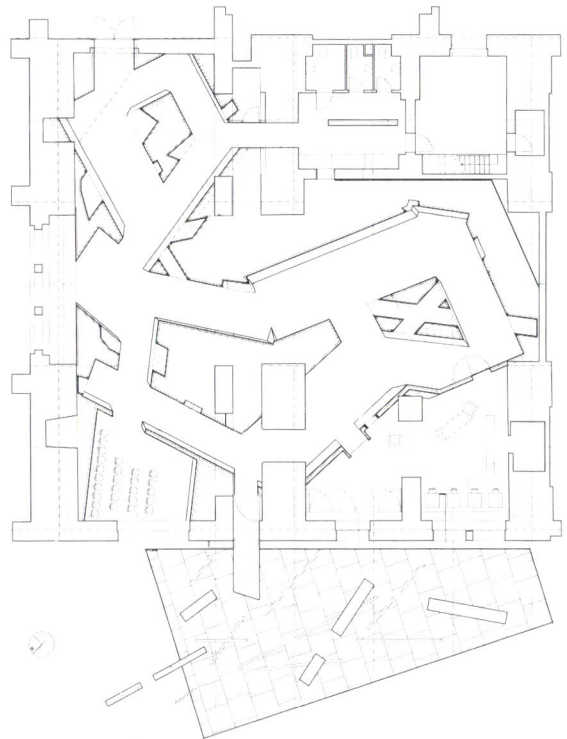

Museum

博物馆

Project Name: Museum   **Designer:** JSª   **Location:** Mexico City   **Photographer:** Jaime Navarro
Time: 2006   **Area:** 345 sqm

项目名称：博物馆 设计师：JSª 设计事务所 项目地点：墨西哥城 摄影：杰姆·纳瓦罗 完成时间：2007年 面积：345平方米

The building known as "La Esmeralda" in historic downtown Mexico City houses Carlos Monsivais' collection. The interior design of the new museum was entrusted to JSª (formerly Higuera + Sanchez). The building's architecture and the collection both are representative of Mexico City's history, and embrace periods from the Colonial times to present day.

How could the designers find the basis for such diverse subjects inside equally different environments?

The designers began with the entrance, where the intention was to capture the spirit of an "estanquillo" (neighborhood grocery shop), in an abstract manner, at the same time seeking to make it the "magic" door that would take the visitor on a trip to the past.

Climbing the staircase, you reach the first floor, which is a large open space, richly adorned with coloured moldings on the ceiling following the design of the cast iron. The furniture in the exposition must be silent, in face of a most eloquent architecture. It is white and formally independent of the environment. Nevertheless, it blends well with the warmth of the ornamental decoration, following soft curves it takes the visitor through the exhibit, where thematic niches are presented occasionally, like closed cells within the space.

此博物馆（被称作 "爱思默瑞达"式建筑）位于墨西哥城著名的闹市区，馆内收集了 "卡洛斯·蒙西瓦伊斯"的作品。 "JSª"（前身是拉伊格拉与桑切斯设计公司）设计公司受到委托，负责馆内设计。博物馆本身以及其内部的作品都体现了墨西哥城的历史以及殖民时代到当今社会的过渡。

设计师们是如何在完全不同的环境里，把众多的主题一一呈现的呢？

首先，设计师们从入口处着手，他们力求使门面看起来像是一家普通的店铺（即 "estanquillo"），同时，又能引领参观者踏上通往历史的时光隧道。

沿着楼梯，参观者到达第一层：整个空间宽敞明亮，其顶棚配有仿生铁设计且着色的脚线。为了突出建筑的特色，展架被设计成白色——形式上独立于整个环境，尽管如此，还会与整个空间的暖格调装饰浑然一体。参观者可以沿着展架优美的弧度参观整个展馆，而间或出现的壁龛，就像是空间内闭合的小室一样。

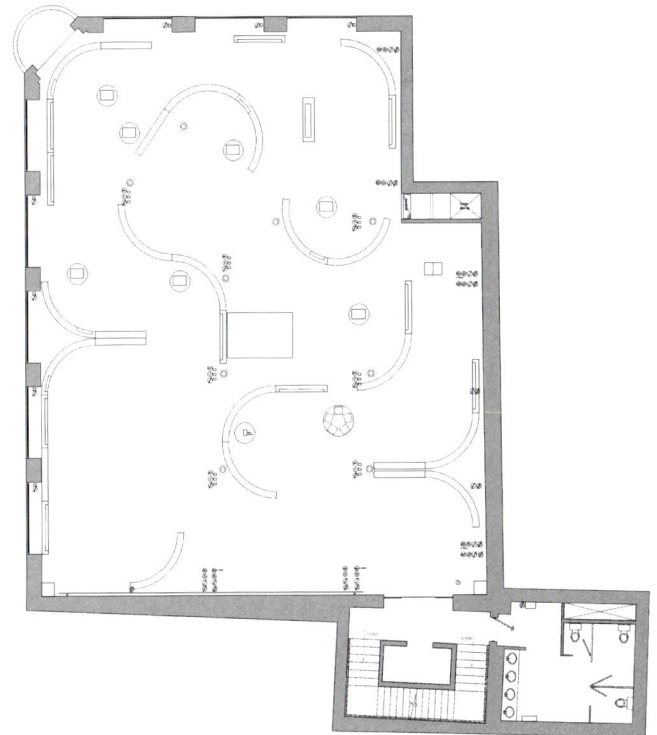

New Acropolis Museum

新卫城博物馆

Project Name: New Acropolis Museum **Designer:** Bernard Tschumi Architects
Location: Athens, Greece **Photographer:** Peter Mauss/Esto, Christian Richters, Bernard Tschumi Architects **Time:** 2007
项目名称：新卫城博物馆 设计师：伯纳德·屈米建筑师事务所 项目地点：希腊，雅典 摄影：彼得·莫斯 克里斯蒂安 里希特斯
完成时间：2007年

Program
With exhibition space of more than 14,000 square metres (150,000 square feet) and a full range of modern visitor amenities, the New Acropolis Museum will tell the complete story of life on the Athenian Acropolis and its surroundings.
Grounds
The Museum is surrounded by 7,000 square metres (75,000 square feet) of landscaped green space.
Amenities
The Museum offers a café overlooking the archeological excavation, a museum store, and a museum restaurant, with a public terrace commanding views of the Acropolis. The Museum is fully accessible to people with physical disabilities.
Principal Design Features
Designed with spare horizontal lines and utmost simplicity, the Museum is deliberately non-monumental, focusing the visitor's attention on extraordinary works of art. With the greatest possible clarity, the design translates programmatic requirements into architecture.
Organisation
The Museum is conceived as a base, a middle zone and a top. The base hovers over the excavation on more than 100 slender concrete pillars. This level contains the lobby, temporary exhibition spaces, museum store, and support facilities.
The middle (which is trapezoidal in plan) is a double-height space that soars to 10 metres (33 feet), accommodating the galleries from the Archaic to the late Roman period. A mezzanine features a bar and restaurant (with a public terrace looking out toward the Acropolis) and multimedia space.
The top is the rectangular, glass-enclosed, skylit Parthenon Gallery, over 7 metres high and with a floor space of over 2,050 square metres (22,100 square ft). It is shifted 23 degrees from the rest of the building to orient it directly toward the Acropolis. Here the building's concrete core, which penetrates upward through all levels, becomes the surface on which the marble sculptures of the Parthenon Frieze are mounted. The core allows natural light to pass down to the Caryatids on the level below.

规划
新卫城博物馆内展区面积超过14000平方米，并设有全套的现代化服务设施，向游客述说着雅典新城的历史以及周围发生的一切。
背景
博物馆的四周环绕着7000平方米的绿色景观。
设施
博物馆内包括咖啡厅、商店、餐厅（带有公共露台，可以欣赏新卫城的壮丽景致）。此外，博物馆符合无障碍设计标准，为行动不便者提供便利。
主要设计特色
设计师将简约理念发挥到极致，大量运用水平线条。他们特意摒弃"永垂不朽"的特色，将游客目光引领到绝妙非凡的艺术品上。此外，设计中最大限度彰显条理性，用建筑形式诠释不同的设计要求。
空间构架
博物馆由三个部分组成：一层、中层及顶层。一层盘旋在100多根精细的水泥梁柱结构上，大厅、临时展区、商店及支持设施设计在此。
中层是一个双层结构，高达10米，容纳着从远古时期到后罗马年代等各个时代的展厅。酒吧、餐厅及多媒体中心设置在两层中间的区域。
顶层呈现长方形格局，高达7米，面积多达2050平方米，全部采用玻璃结构打造，帕特农神庙展厅便坐落在此。这一层偏转建筑其他部分23度，直接朝向卫城。

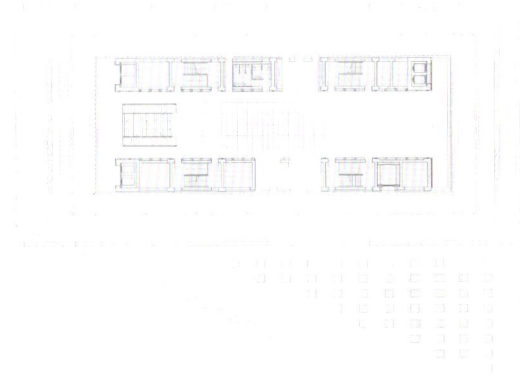

The Art Institute of Chicago — The Modern Wing, Chicago "现代之翼"博物馆

Project Name: The Art Institute of Chicago—The Modern Wing, Chicago **Designer:** Renzo Piano Building Workshop, Paris, France **Location:** Chicago, USA **Photographer:** Dave Jordano **Time:** 2009

项目名称："现代之翼"博物馆 设计师：伦佐·皮亚诺 项目地点：美国，芝加哥 摄影：戴夫·乔丹诺 完成时间：2009年

The new Modern Wing is being built between Michigan Avenue and Columbus Drive, at the northeast corner of the block the Art Institute of Chicago currently occupies. The addition will complete the cultural, urban campus of the museum. The new street-level entrance on Monroe Street will connect Millennium Park to the heart of the existing museum through the new Griffin Court. On the first floor, this daylit court will be flanked by new educational facilities, public amenities, galleries, and a garden, all of which will better actively link the Art Institute with urban life. The second and third floors will be dedicated to art and the viewing of art. The third floor will be completely lit by natural light. Below street level will be mechanical systems, art storage, and support facilities for the entire Art Institute.

Flying above the art pavilion will be a shelter that filters the sun to create the natural shaded light conditions ideal for the enjoyment of art. This shelter is a kind of flying carpet made of aluminum leaves that perform the same job as the tree canopies all around in the park. It is a "soft machine" that sensitively levitates above the new wing, vibrantly screening the light.

Limestone, a material used in the construction of the entire museum from its original Beaux Arts palace to recent additions, rises from the ground like a topographic relief, massive and solid, as though it has always been there. Above this topos, the building stands light, transparent, and permeable in steel and glass, in the great tradition of Chicago buildings: solid and robust yet at the same time light and crisp.

新建"现代之翼"博物馆位于密歇根大道及哥伦布私人车道之间——芝加哥艺术学院街区东北角。新建结构更加完善了博物馆的文化及城市园区，位于门罗街一侧的入口设于界面处，将世纪公园同原有博物馆中心结构连通。一层光线充裕，包括教育机构、公共空间、画廊及花园；二层和三层主要用于展示艺术作品，其中三层完全采用自然光线照明。地下一层主要用于存储机械设备、艺术品及支持设施等。

艺术展厅之上悬浮着遮蔽结构，其功能即为将光线过滤以营造适合艺术品展出的完美环境。这一结构主要由铝箔材质打造，好似一块漂浮的地毯。

石灰石材质在整个博物馆内随处可见，从地面处"拔地而起"、蜿蜒起伏，好似同空间与生俱来一般，凸显出厚重结实之感。钢材和玻璃材质的运用则传达出轻盈、通透之感。分明的对比更是进一步彰示了芝加哥风格建筑的伟大传统：稳固动感但不乏轻盈利落。

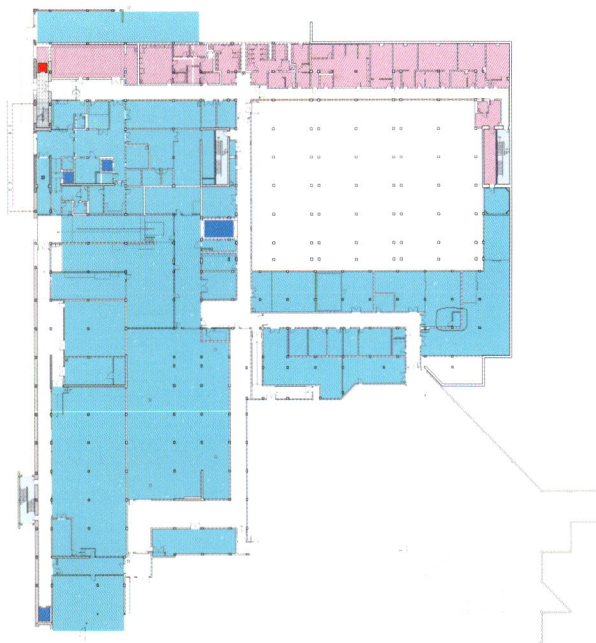

1

Arts of Asia

The Arts of Asia Gallery, Auckland

亚洲艺术画廊

Project Name: The Arts of Asia Gallery, Auckland   **Designer:** Catherine Stormont, Nicole Pfoser, Gustavo Thiermann **Location:** Auckland, New Zealand   **Photographer:** HBO+EMTB   **Time:** 2008

项目名称：亚洲艺术画廊 设计师：凯瑟琳·斯托蒙特 尼科尔·普弗瑟 古塔斯沃·泰尔曼 项目地点：新西兰，奥克兰
摄影：HBO+EMTB建筑咨询公司 完成时间：2008年

The gallery is located in an existing exhibition space at the Auckland War Memorial Museum. The minimalist but finely detailed gallery showcases over 300 artefacts, providing built form that does not compete with the artefacts. The design team worked closely with the Gallery's curator to determine the most appropriate manner in which individual items are best presented, with consideration for chronology or theme.

Part of the challenge for the design and museum teams was to reduce it down to a manageable size to fit the exhibition space and still tell the story in a chronological sequence. The display cases need to sit within this space as a separate element but also needed to unify the space.

The concept for the gallery was to create a clean, linear and simple space which showcased the various pieces at their best while still keeping the valuable objects safe. In addition it was important to allow the audience maximum access to the pieces and for them to be able to get a close view if needed. Many of the pieces include intricate detailed work while others are large and need to be viewed from afar.

In order to enhance this it was decided that the cases should be constructed largely of glass creating an "invisible" barrier between the audience and the objects. Imperative to the aesthetic of the design was the visibility of the objects in the display cases. As few as possible "interruptions" in sight lines were needed which required the use of large glass panes which meant fewer joins. Innovative ideas were needed to support these large structures while still allowing the curators easy access to the objects for maintenance purposes.

Careful use of lighting was used to enhance the sense of drama and intimacy.

The desired aesthetic was to create a calm, tranquil space that allows the viewer to enjoy the individual beauty of the decorative objects while seeing them in the context of a historical timeline.

亚洲艺术画廊位于奥克兰战争纪念博物馆内，其内展出的艺术品逾300件。设计团队同馆长共同协作，努力使每一件作品都能得以最好的展示，以年代或主题为顺序。

设计面临的挑战便是如何将展示空间缩减到合适的大小，但同时不影响作品的展出顺序。展台作为单独的结构，更起到统一空间的作用。最终理念是营造一个整洁、简约的线性空间，让作品得到充分展示并确保贵重展品的安全。另外，还需考虑让游客近距离接近展品，而那些大型的艺术品则只能远观。

为满足这一要求，设计师专门打造了玻璃展柜，在游客和展品之间建立起一道无形的屏障。保证展品以最真实的面貌呈现是重中之重，因此要尽量减少展柜之间的连接点。另外，设计师更构思了极具创意的办法维护展柜不受损害，同时确保管理员们随时打开以维护保养内部的展品。

灯光更是经过精心打造，突出空间艺术感和亲切感。游客们在淡雅幽静的空间内品味着每件艺术品特有的美感，并将其与特定的历史背景联系起来。

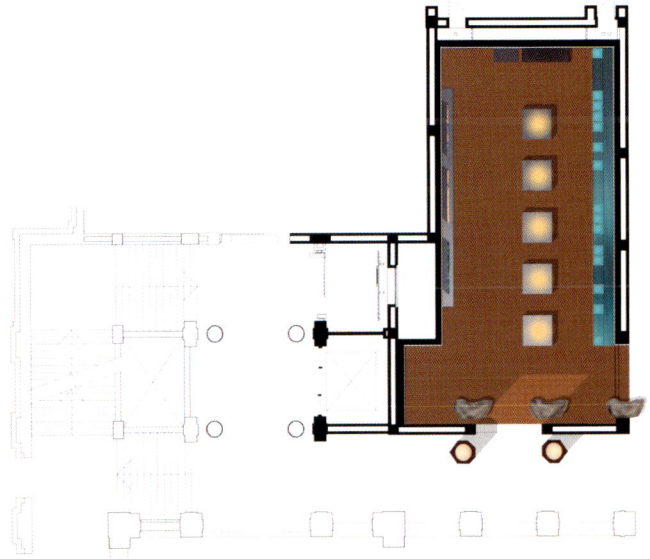

That's Opera

"那是" 剧院

Project Name: That's Opera　**Designer:** Atelier Brueckner GmbH　**Location:** Brussels, Belgium
Photographer: Atelier Brueckner GmbH　**Time:** 2006

项目名称："那是"剧院　设计师：布鲁克纳工作室　项目地点：比利时，布鲁塞尔　摄影：布鲁克纳工作室　完成时间：2006年

200 Years of Italian Music
Temporary Exhibition

The opera is in the limelight. In celebration of the 200th anniversary of the music publishing company Ricordi, the opera can be experienced from an unusual point of view. A red carpet takes the visitors on a journey to the theatre's backstage world. They interactively experience the genesis of an opera as a "Gesamtkunstwerk," or synthesis of the arts, step-by-step in five exhibition cubes. They are present backstage when Libretto and Partitura find their written expression, when stage design and costumes take form and eventually come on stage for the performance.

Individually designed spatial narratives put the visitors in characteristic sceneries, for example, Rodolfo's garret – inspired by the opera "La Bohème" – becomes the scriptorium for the librettists in the first exhibition cube. In the following cube "Partitura", the visitors enter an orchestra pit, where the instrumental parts are distributed on 42 directional loudspeakers. The cube "Scenografia" takes them to the workshop of a stage designer, while the setting of "Madama Butterfly" inspires the cube "Voci e Costumi". The latter is lined with backlit Shoji wall elements reminiscent of Japanese architecture. At the end of their tour, the visitors enter La Scala Theatre of Milan. Here, a 270-degree projection surrounds the visitors with a contemporary staging of Aida: They can experience the opera from varied and unusual perspectives, including from the stage.

The exhibition architecture follows the aesthetics of stage setting construction as well as the materiality and degree of backdrop detail. The visitors are backstage in rooms that are normally not accessible to them. Supports, wiring and spotlights are visible and arouse their curiosity of the cubes' interior. These cubes are composed of one-metre-wide by three-metres-high elements, which are shrouded with different materials. Thus, they can be adapted to every location. The modular construction method allows for flexibility.

意大利音乐200年临时展

为庆祝意大利音乐发行公司——里科尔迪成立200周年，"那是"剧院再一次成为这座城市的焦点，从一个非同寻常的角度诠释着自身。来访者踩着大红地毯一直走到剧院后台，流连在5个展厅内，体验着艺术诞生的旅程。

展厅设计各具空间特色，让游客在不同的"背景中"往返。第一个展厅中"鲁道夫阁楼"空间受到La Bohème剧院的影响，为作词者们提供了一个灵感之所。紧接着便是Partitura展厅，游客走进乐池，乐声从42个定向扬声器中飘散出来，令人沉醉。Scenografia展厅带领大家走进舞台设计者的"工作室"，Voci e Costumi展厅的设计灵感源于歌剧《蝴蝶夫人》的布景，背光照明的障子结构，让人不禁联想到日式建筑。最后，游客们走进"斯卡拉剧院展厅"，三面环绕的巨大投影仪上不断播放着《阿依达》，让人从不同的角度欣赏着这部佳作。

另外，展厅建筑追随着舞台设计风格，同时强调功能性及细节。游客们可以进入平时很难走进的后台，伴奏音乐、聚光灯无一不点燃他们的好奇心。所有展厅全部采用3x1米的模块结构打造，上面被不同材质包裹起来。这一设计大大增强了灵活性，便于展厅从一个地点移动到下一个场所。

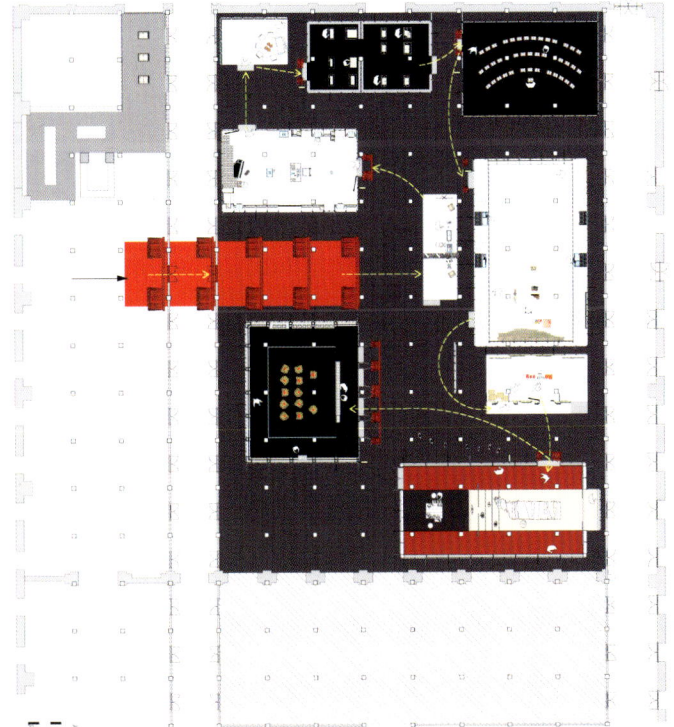

Piazza della Scala in the first half of the 20th century

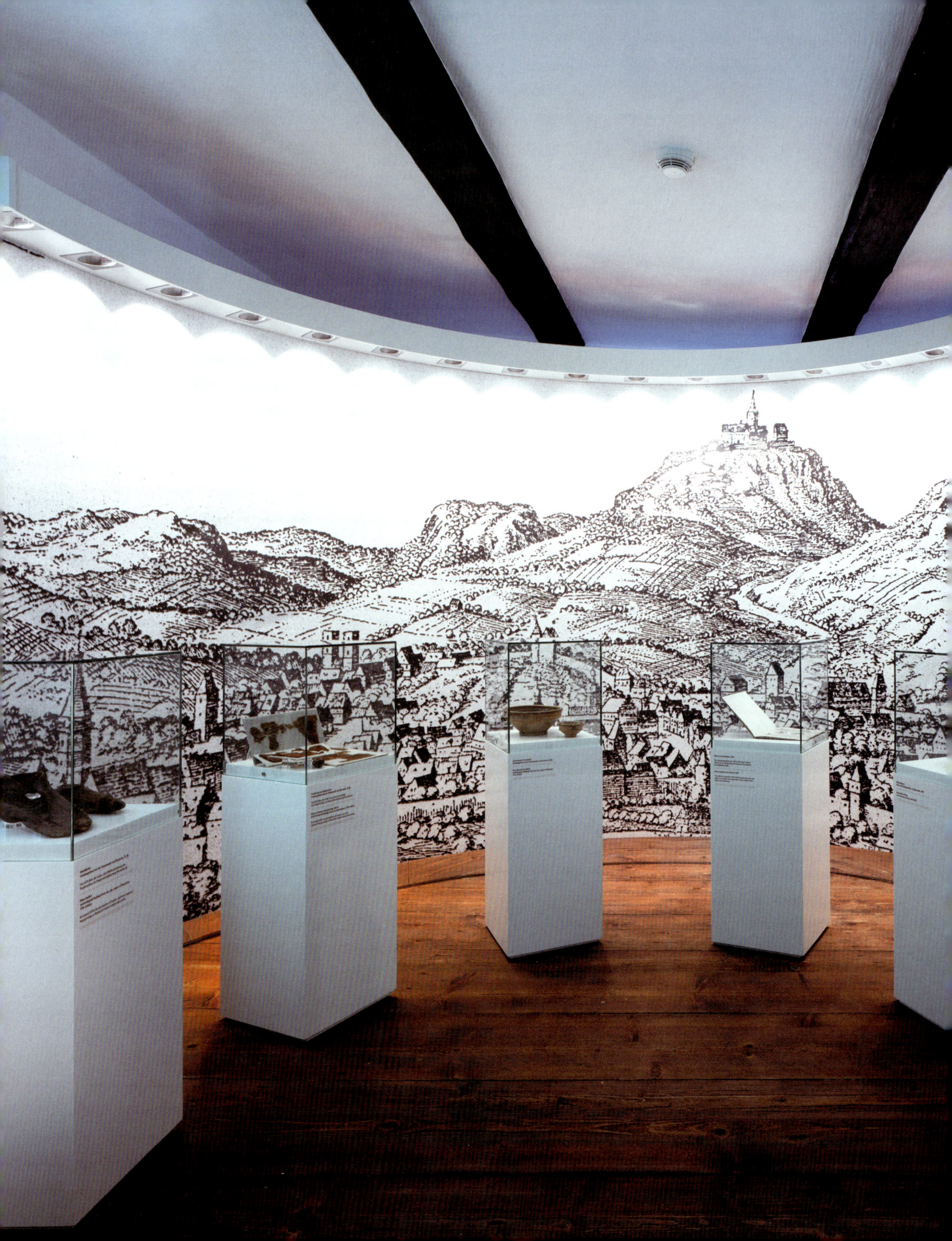

Bachhaus Eisenach

巴赫纪念馆

Project Name: Bachhaus Eisenach **Designer:** Penkhues Architekten, Kassel **Location:** Eisenach Germany **Photographer:** : Penkhues Architekten, Kassel **Time:** 2007

项目名称：巴赫纪念馆 设计师：Penkhues建筑师事务所 项目地点：德国，爱森纳赫 摄影：Penkhues建筑师事务所 完成时间：2007年

The Bach House in Eisenach, established in 1907 as a memorial for the composer Johann Sebastian Bach (1685-1750), is the starting point for admiration and research of Bach worldwide. The renovated and architecturally supplemented building ensemble was anew opened to the jubilee on the 17th May 2007. A new permanent exhibition, implemented by Atelier Bruckner in Stuttgart, puts the emotional, mental as well as musical potentials in Bach's work in the centre of the presentation. For the first time the sensuous experience of his creating is possible beside the mediation of biographic, historical and musical knowledge. A new building (after the plans of Penkhues Architekten, Kassel) offers the possibility to make the musical work of Johann Sebastian Bach learnable in its complexity and greatness. The new building closes a gap in the construction ensemble which exists of old building with cultivations, forecourt and bipartite court situation.

A diversified walk through the Bach House was established for the visitor. Through the entrance hall of the new building he reaches the historical old building where the history of the Bachgesellschaft, bearer and founder of the museum, is introduced and a music hall invites the visitors to linger. Valuable historical music instruments are not only shown here, but are explained and played every hour.

The upper floor of the old building is dedicated to the life of Bach in his historical environment. The new building provides the newest researches to the work of Johann Sebastian Bach and makes the musical experience very vivid. The individual elements of the exhibition design are employed very accurately, broken down and precise. Like a fugue of Bach, basic elements of the same kind in different variations are forming the spatial view. Referring to that idea only the materials aluminium and matt finished acrylic glass are employed throughout the exhibition.

巴赫故居始建于1907年,用于纪念伟大作曲家同时开创了世人对巴赫的崇拜及研究的起点。2007年,设计师将原有建筑翻新并扩建,从精神、智慧以及音乐天赋等方面全方位展现巴赫及其作品。新楼填补了旧建筑的空缺,主要用于展现其音乐作品。穿过新楼的入口大厅,游客便可达到老楼,了解纪念馆的创始人的历史,同时还可观赏到价值连城的乐器以及聆听每个时段的演奏。

旧楼的上层用于介绍巴赫的生平历史,而新楼则着重展现其创作,重现巴赫的美妙音乐。展厅设计中精确运用每一个细节,铝材及亚光玻璃的运用旨在突出展品。

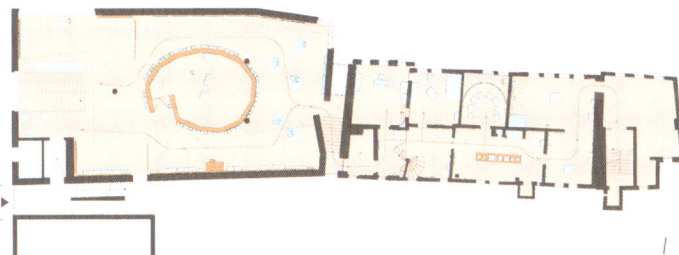

Weimar, Residenz des Herzogtums Sachsen-Weimar, stand unter der doppelten Regentschaft der Herzöge Wilhelm Ernst und Ernst August. Als Bach 1708 von Mühlhausen aus die Stelle des Hoforganisten und Kammermusikers antrat, geriet er in die Intrigen zwischen den Herzögen. Als Mitglied der Hofkapelle unterstand er dem jungen religiösen Herzog Wilhelm Ernst, doch der junge Herzog Ernst August, selbst Violinist und Trompetenspieler, liebte und förderte ihn und machte ihn zum Musikleiter seines jüngeren Mühlhausen.

Bachs Hauptaufgabe war die des Hoforganisten – aus dieser Zeit stammt das Orgel-Büchlein. Im Frühjahr 1714 avancierte Bach zum Konzertmeister und spielte damit eine führende Rolle bei Aufführungen der Hofkapelle – in Konkurrenz zum Kapellmeister und Vizekapellmeister. Sein Vertrag verpflichtete ihn dazu "monatlich neue Stücke" für den Gottesdienst in der Schloßkirche zu schreiben.

In Bachs Weimarer Zeit fällt seine intensive Auseinandersetzung mit dem Konzertstil Vivaldis. Zu den hier entstandenen Werken gehören die Jagdkantate, die Erstfassung des 1. Brandenburgischen Konzerts und sohl die Violinsonaten. Nur ca. 15 – 20 Prozent von Bachs Weimarer Kompositionen sind erhalten.

Weimar, the capital of the Duchy of Saxony-Weimar, was ruled jointly by Duke Wilhelm Ernst and Duke Ernst August. When Bach took up the post of Court Organist and chamber musician from Mühlhausen, he became involved in the intrigues between the Dukes, as a member of the Court orchestra, he was answerable to the strictly religious Duke Wilhelm Ernst, but the young Duke Ernst August, who was himself a violinist and trumpet player, loved and encouraged him, and appointed him music teacher to his younger half-brother.

Bach's main work was that of the Court Organist – the Orgel-Büchlein dates from this time. In the spring of 1714 Bach was promoted to Concert-Master, which meant that he played a leading role in performances in the Court orchestra – in competition with the Director of Music. His contract obliged him to write "new pieces each month" by the service in the Palace Church.

During Bach's time in Weimar, he made an intensive study of Vivaldi's concerto style. The works that he wrote during this period include the Hunt Cantata, the early version of the 1st Brandenburg Concerto, and probably the violin sonatas. Only around 15 – 20 percent of Bach's Weimar compositions have survived.

Herzog Wilhelm Ernst (1662–1728)
Kupferstich, nach 1690

Herzog Wilhelm Ernst von Sachsen-Weimar (um 1728 – 1733)

Ernst Wilhelm Ernst (1662–1728)
Duke Wilhelm Ernst von Sachsen-Weimar (um 1728 – 1733)

Herzog Ernst August (1688–1748)
Kupferstich von Johann Christoph Sysang, 1742
Original: Bachhaus Eisenach

Herzog Ernst August von Sachsen-Jena/Sachsen-Weimar in Weimar

Duke Ernst August von Bach's medal appears in Weimar

Multikino Szczecin

玛尔缇凯诺什切青电影院

Project Name: Multikino Szczecin　**Designer:** Robert Majkut Design studio　**Location:** Szczecin, Poland　**Photographer:** : Maciej Frydrysiak　**Time:** 2006

项目名称：玛尔缇凯诺什切青电影院　设计师：罗伯特玛塞杰设计工作室　项目地点：波兰.什切青　摄影：玛塞杰·弗雷德雷塞克　完成时间：2006年

The project was to introduce Multikino to obtain cinema complex with unique and attractive character. The place which gives the viewers impression of peculiarity even before one enters the cinema hall.

Half-round and wide cash boxes perfectly introduced into the whole interior are one of the obvious stages. The assumption of main hall, particularly front well, was presenting the place of the cinema location as a part of the universal "world of the film".

Therefore, photographs of people resembling which are taken by paparazzi pictures of film stars are placed against a background of Earth, seen from the outer space. The viewer can not only watch a film here or stay in the interesting interiors, but also feel as a part of a film world, vibrant with activity, not only abroad in Los Angeles, London or New York, but in Szczecin as well.

The design includes a strong colour science fiction rather a sense of the future, as the audience entered the science fiction movies. The main colours of red, impressive and easy to create the feeling of excitement and happiness. The roof is black, the large spherical chandelier hanging in the air, like the shining stars in the night.

该项目的主旨是将玛尔缇凯诺打造成独特又引人入胜的电影院。独特的选址让观众在踏入影厅之前就能领略其新奇之处。

入口处开阔的半圆式电子收银系统的设计醒目，巧妙地将来宾引入室内空间。正厅，尤其是其前端，设计别致，令观众好似步入一个广阔的电影世界。

在地球形状的背景下张贴着酷似影星的照片，球形背景好似从外太空看到的一样。身处一个颇具情趣装饰的影院空间，观众不仅可以欣赏一场电影，还可感受到广阔的电影世界所带来的震撼与愉悦，这种体验不仅存在于国外的洛杉矶、伦敦和纽约，在什切青同样可以感受到。

该影院的设计带有浓浓的科幻色彩，颇具未来感，如同走进了科幻电影的画面之中。选红色为主色调，醒目且更令人兴奋和快乐。顶棚选用黑色，大大的球形吊灯悬于半空，恰似出现在夜幕之中的星星，闪闪发亮。

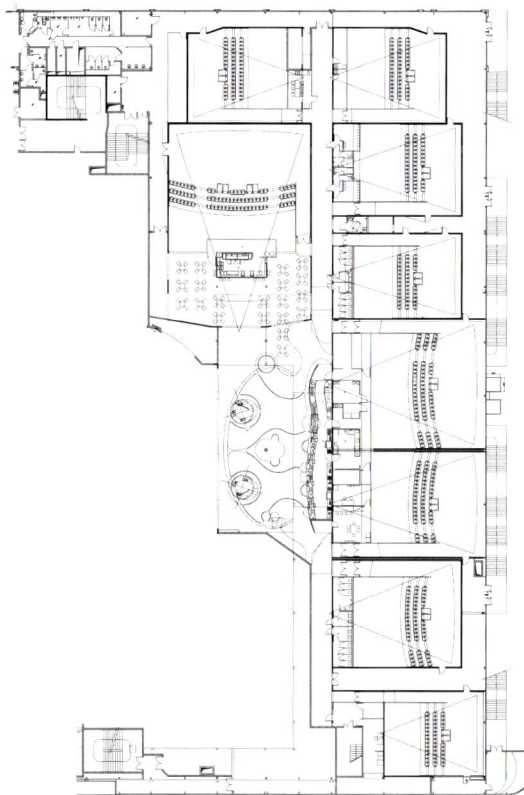

Blue Room Theatre at Chesapeake

宝龙影院

Project Name: Blue Room Theatre at Chesapeake **Designer:** Elliott+Associates Architects
Location: Oklahoma City, USA **Photographer:** Scott McDonald, Hedrich Blessing **Time:** 2007

项目名称：宝龙影院 设计师：埃利奥特联合建筑师事务所 项目地点：美国，俄克拉荷马城 摄影：斯科特·麦克唐纳、
海德里奇·布莱辛 完成时间：2007年

The Blue Room is an element of Chesapeake Energy CEO Aubrey McClendon's overall goal to attract the corporation's next generation of accountants, engineers, geologists and technicians. McClendon's objective was to provide a multi-faceted centre that would enhance the workplace environment and provide greater employee benefit and retention.

The three-year design and development process of the Blue Room Theatre fulfilled this objective, culminating with a high end, cutting edge facility. The Blue Room Theatre is used for Chesapeake Energy's corporate meetings, videoconferencing, IT and operations functions, and geology and engineering department lectures. The Blue Room Theatre also serves as a movie theatre for employees during off-work hours.

The design takes advantage of the ascension/subterranean location to enhance an unexpected spatial experience; light and colour to create an immersive atmosphere; reflective, transparent, translucent and absorptive materials to create a superb acoustical environment; sound reflective blue cellular polycarbonate panel side walls which glow from backlighting; ceiling mounted MR 16 lamps containing clear and blue filters to provide white light for writing and blue light for atmosphere; dimmable room lighting settings for all white light, all blue light, combined colour and movie mode; and blue internal fiber optic glass column lighting.

宝龙影院是切萨皮克能源公司整体规划的一部分，旨在完善员工的工作环境，让他们享受到更多的优势。

设计历经三年时间，最终因高档环境及先进的设备而备受欢迎。影院的主要功能是举办公司会议、视频会议、IT测试、地质及工程部门演讲等活动，同时在业余时间为员工播放电影，丰富他们的生活。

设计充分运用了影院内逐渐升高的地势，增强空间感。光线及色彩营造了独特的氛围；透明及半透明吸音材质的运用打造完美音质效果背景；蓝色蜂窝式聚碳酸酯板材墙壁在光线反射下闪烁着；天花板上安装的灯饰包含白光及蓝光过滤器（为字幕提供白光，为其他空间提供蓝光），别具特色。

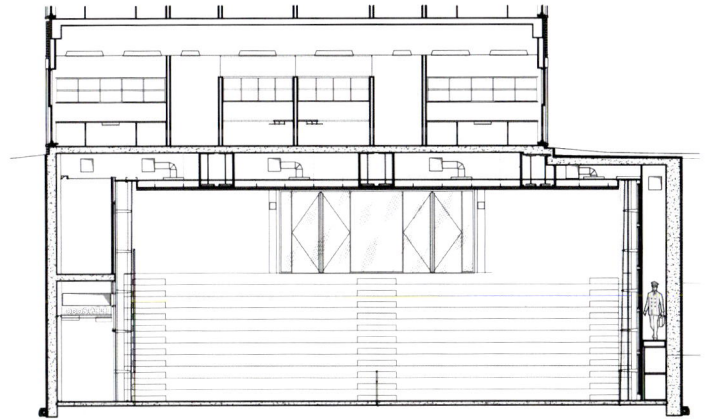

Ordrup School

奥竹普学校

Project Name: Ordrup School   **Designer:** Bosch & Fjord   **Location:** Copenhagen Denmark
Photographer: Anders Sune Berg   **Time:** 2006

项目名称：奥竹普学校 设计师：博世&弗乔德事务所 项目地点：丹麦，哥本哈根 摄影：博世&弗乔德事务所 完成时间：2006年

Based on the conviction that all people are different and have different ways of thinking and learning, at Ordrup School in Gentofte, Denmark, Bosch & Fjord have rejected the traditional design of school interiors and instead created varied rooms with space for differentiated teaching and creative thinking. The design is based on three key concepts: "peace & absorption", "discussion & cooperation" and "security & presence".

In the younger classes, peace and absorption are emphasised in the upholstered reading tubes, while movable pieces of carpet create temporary spaces for discussion and cooperation. In the mid-range age levels, the students can work together in smaller forums inside the sculptural Hot Pots or withdraw to the colourful concentration booths for concentrated reading and other work activities without being distracted by the surroundings. In the oldest classes, importance has been attached to the teenager's situation of being "on the way out into the real world". The bright red sofa islands on wheels can be moved around on the bright yellow floors and used for concentrated group work, loud discussions or movie showings. A long, bright green table sprawls through one of the rooms, forming a dynamic setting for creative cooperation and flexible work situations.

奥竹普学校位于丹麦，设计师基于"思维方式及学习方法因人而异"的理念，在室内设计上打破传统样式，打造不同的空间，便于培养学生的创意思维。设计围绕着三个主要理念展开——"平和与专注"、"讨论与合作"、"安全与展示"。

但低年级教室中，设计主要强调"平和与专注"。移动的坐垫打造了临时的座椅，便于孩子们相互讨论与交流。中年级教室中，学生们可以在"暖水瓶"中举办小型论坛会议，或是到彩色小亭中专注于自己的阅读。高年级的教室中，在设计上旨在为他们营造一个真实的世界。鲜红色的沙发安装着轮子，可以在黄色区域内自由移动，便于学生们的团体工作以及大声讨论。

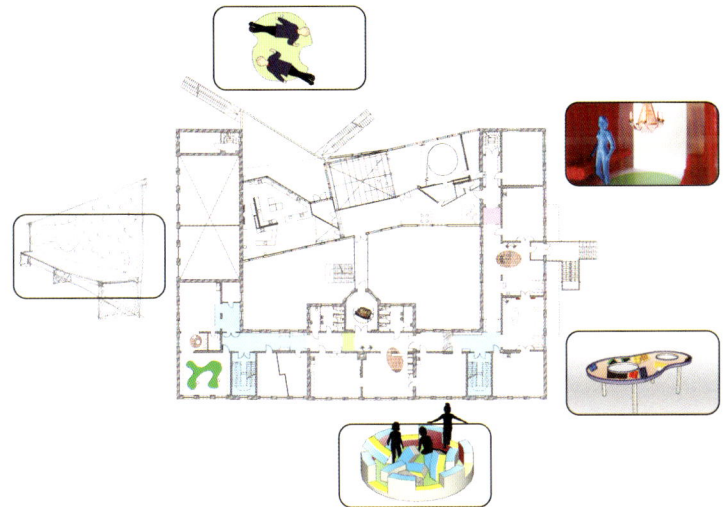

Medical Faculty, University of Groningen

格罗宁根大学医学部

Project Name: Medical Faculty, University of Groningen **Designer:** RAU
Location: Groningen, The Netherlands **Photographer:** Ben Vulkers Bjorn Utpott **Time:** 2008

项目名称：格罗宁根大学医学部 设计师：RAU 项目地点：荷兰，格罗宁根 摄影：本·伍克斯 比约恩·尤特波特
完成时间：2008年

Joined Movement

The Orthopaedic Institute wanted to erect a major building here to house its institute's rooms and a main lecture hall with space for 450 students and with a large foyer, connected to the institute's existing rooms. The new building would therefore have a "jointing" function in urban and programmatic terms as well.

Behind the glass revolving doors of the entrance area, facing the university plaza, an airy space is organised into 3 storeys developed, whose side walls partially consisted of the external walls of the existing developments. From the foyer, a single staircase leads to the two upper storeys, which house all the institutional rooms for orthopaedics. Within its two double arms, the first floor encircles an inner courtyard with garden, laid out on the roof of the lecture hall.

At the front of the foyer, a creatively designed concrete wall rises, which is the highest point of the giant lecture hall, half sunken into the earth. This wall, "Chromosome Technology in Concrete Relief", was designed in close collaboration with the artist Baukje Trenning. The lecture hall itself falls away towards a large white projection screen. Tables and chairs are combined to form a single large unit of furniture. They are sprayed red all over, a surface treatment technique borrowed from the motor industry.

A Skin of Gold

At the centre of the building is the concrete skeleton, a self-supporting column and ceiling construction. Steel "muscles" are connected to this "skeleton". This is covered by a gleaming golden "shawl", which spans the building. The double-curved aluminium "skin" melted so as to flow, involved a special craft technique. Multiply curving surfaces are not unusual on aeroplanes and ships. For this reason, a shipbuilder was recruited to deal with the statics. All of the double-curved surfaces which are also visible from the inside were made as prefabricated parts. The overlapping of the façade panels provides natural inlets fresh air.

连通结构

格罗宁根大学医学部要求打造一个全新的建筑，用于研究室和演讲大厅，供450名学生使用。宽阔的前厅与原有空间相通，因此这一新结构便起到了"连通"功能。入口处高大的玻璃旋转门后是一个三层的通风结构，部分侧墙借用了原有结构的外墙。楼梯从门厅处延伸，一直通往上面两层。

另外，一面创意十足的混凝土墙壁"矗立于"前厅的最前面，一半"陷于"地下构成了演讲大厅的最高点。演讲大厅内安装有巨大的白色投影幕布，桌椅全部漆成红色（采用汽车工业专有面漆处理技术），摆放在一起构成了独立的家具结构。

金色表皮

建筑的中心结构即为混凝土"骨架"，钢材则如同"肌肉"一般紧紧贴合在上。除此之外，最为引人注目的便属金色"表皮"（双曲铝制表皮采用特殊工艺融合在一起，这一技艺只用于飞机及船舶制造领域）。此外，通过打孔技术处理，从建筑内部同样可以看到独特的外观，当然自然清新的空气同样可以渗透进来。

Chesapeake Fitness Centre, East Addition

切萨皮克健身中心扩建

Project Name: Chesapeake Fitness Centre, East Addition   **Designer:** Elliott + Associates Architects
Location: Oklahoma City USA   **Photographer:** Scott McDonald Hedrich Blessing   **Time:** 2009

项目名称：切萨皮克健身中心扩建 设计师：埃利奥特联合建筑师事务所 项目地点：美国，俄克拉荷马城
摄影：斯科特·麦克唐纳 海德里奇·布莱辛 完成时间：2009年

The Chesapeake Fitness Centre, East Addition is located on the campus of the Chesapeake Energy Corporation. The first Fitness Centre was constructed in 2003 and this is an expansion to that building. Headquartered in Oklahoma City, the company is the second-largest producer of natural gas in the nation and the most active driller of new wells in the U.S. The company's commitment is to encourage employees to live a healthy lifestyle, which is reflected in nearly every aspect of the campus. The fitness center, a 7,145,000-square-foot facility, embodies this by offering employees the ease of keeping fit without having to leave campus. Aerobics, cardio training, swimming, fencing, weightlifting, basketball and cycling are just a few of the programs offered under the guidance of an experienced fitness staff.

The designed addition is based on 'Jewel next to creek' concept. It is 1,450,906 square feet. The program requires that first, there should be three workout rooms. Second, there should be one regulation squash court. Third, there should be one additional basketball court with workout loft and last of all, the men's locker room should be enlarged.

Architectural Concept:

First, the glass workout tower with workout areas are transparent to campus but not to the public road.

Second, emphasising movement – people stretching in glass walkway and going up and down stairs are visible from public road.

Third, there is an articulate difference between structure and skin – glass is pulled away from concrete structure. The glass workout tower gives the feeling of being up in a tree house next to the creek.

切萨皮克健身中心位于切萨皮克能源公司(美国第二大天然气供应商)园区内，原有建筑始建于2003年，占地6689平方米，现决定对其进行扩建。切萨皮克能源公司一直鼓励员工健康生活，健身中心的建造正是符合这一理念。

扩建空间占地1942平方米，设计理念源于"河边的宝石"(意指比原来的结构更加出色)。这一工程包括四个部分:新建三个室内训练空间;新建一个壁球场; 增建一个篮球场(带有训练场地);扩建男士衣物间。

建筑理念

建筑理念由三个部分构成:

首先,玻璃结构训练场馆需朝向园区一侧完全开放,朝向道路一侧则无需此要求;

其次,强调动态感——人们在训练馆内来回走动的场景在道路上能一目了然;

最后,内部结构同外观之间应区分开来——玻璃外观与混凝土结构分离。

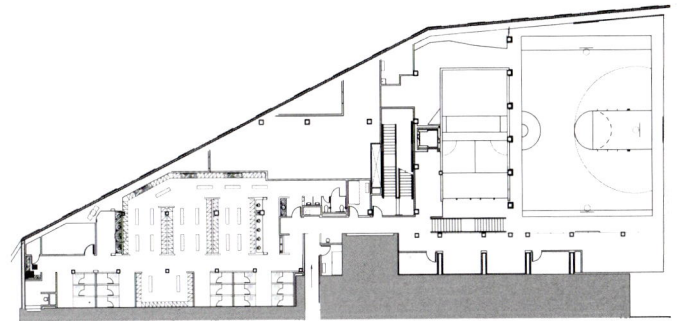

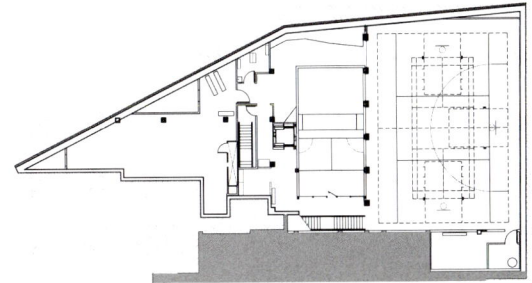

14 Street Y Renovation

14号街

Project Name: 14 Street Y Renovation   Designer: Studio ST Architects   Location: New York USA
Photographer: Bilyana   Time: 2009

项目名称：14号街 设计师：ST建筑师工作室 项目地点：美国，纽约 摄影：比亚纳·季米特洛娃 完成时间：2009年

On the ground floor, the lobby, fitness centre, locker rooms, office, showers and pool are reorganised as a series of parallel bands. As users move through the different spaces, they experience the variety of happenings that animate the building.

The existing offices, which occupied the prime space along 14th Street, were relocated to make space for a new entrance lobby. This 469000-square-foot lobby has a gradient blue custom pattern cement tile floor, a field of circular fluorescent lights, and bright yellow 100% recycled plastic lounge furniture. New large windows were installed that open the lobby to 14th Street and bring the action of the street and natural light from outside into the space. A new garden was planted with colored pebble and birch trees to shade the facade. The overall effect is young, fresh and befitting the Y's East Village location.

The fitness centre, while not much larger than it was originally, now comfortably accommodates more fitness equipment. This space is also organised as a series of parallel bands corresponding to type of equipment (cardio, weight training, stretching etc.). The bright yellow and orange rubber floor brighten this windowless space. The waiting bench, fitness desk and low partition walls are all formed of curved rubber.

Locker rooms were relocated to the southern end of the fitness centre. The architects removed the existing vinyl floor and poured a clear layer of epoxy mixed with sand to create a slip-resistant concrete floor. The warm brown colour of this floor is juxtaposed with six vibrant colours of locker doors. Each colour indicates the type of locker, from monthly rentals to day use and handicap-accessible. The light fixtures, which are typical of construction sites and sidewalk scaffolding, float below the exposed concrete ceiling, creating a simple plane that complements the existing pipes and ducts servicing the upper floors.

一层包括大厅、健身中心、衣物间、办公室、淋浴区及游泳池,这些区域全部经过重新规划。客人们在不同的区域走动感受着这里发生的一切。

原来沿着14号大街一侧、占据黄金地点的办公空间被移除,旨在为入口大厅"让位"。新建大厅面积达186平方米,地面铺设着蓝色瓷砖。圆形荧光灯饰以及亮黄色再生塑料家具更具特色,宽敞的大窗朝向大街,让光线洒落进来。

健身中心空间大小未经太大改动,一系列平行设置区域专门为摆放各种器材而打造。鲜黄色及橘色的橡胶地板使得无窗户的空间更加明亮,等候座区、登记处及低矮隔断墙全部由弯曲的橡胶材质打造。

衣物间搬迁到健身中心的最南端,原有的塑料地板被拆除,重新铺设了防滑水泥地面。暖褐色的地面与六色衣柜门形成完美组合,灯具悬浮在裸露的天花板上,形成了一个简单的平面。

Y+ Yoga Centre

Y+瑜伽中心

Project Name: Y+ Yoga Centre   **Designer:** Lyndon Neri and Rossana Hu   **Location:** Shanghai, China   **Photographer:** Derryck Menere   **Time:** 2006

项目名称：Y+瑜伽中心 设计师：林登·奈里 胡如珊 项目地点：中国，上海 摄影：德里克·梅奈莱 完成时间：2006年

Y+ Yoga and Wellness Centre is an extension of the original Y+ Yoga Studio. The total area for this multi-purpose wellness centre is approximately 1200 m². The Centre, opened in December of 2005, contains three yoga rooms, massage rooms, meditation rooms, social gathering spaces, a small café, and a retail space.

The two internal yoga rooms have leaf patterns randomly placed on the walls and ceiling. One room has protruding light fixtures and the other has recessed light fixtures, creating a linear directional experience that emphasises fluidity. The third yoga room is a slightly elevated half-circular room in an abstracted forest clearing looking through the windows at a lake. The forest is abstracted with vertically hung ropes dyed in different shades of green representing a series of abstracted trees.

Additional rooms as part of the design serve to offer a complete oasis. The white room is an entirely white space from the ropes which hang from the ceiling, to the walls, floors, and cushions. The absence of colour allows an alternative space for self-reflection. The rooms for encounter are intermediary rooms that are strategically placed between the circulation and the yoga rooms. These spaces are for intentional and accidental encounters. Encounters between people are an essential part of human existence. These include rooms to cool down, rooms for reading, rooms to chat, rooms to settle down and rooms for meeting new people.

The bronze gaze holes, placed strategically throughout the space, allow the yoga performer to focus on a hole for various yoga positioning or posturing. From the corridors, the portholes also indulge the voyeuristic nature in all of us to gaze into a world of tranquility. This absorption of the world of tranquility into the "self" is initiated by the gaze-perception-sweeping out into the world.

Y+瑜伽中心位于上海新天地，是一个可以容纳1200人的多功能健身中心，创建于2005年12月，包括三个瑜伽室、按摩室、冥想室、交流区、一个小咖啡厅和一间零售店。

三间瑜伽房中，其中两个房间里树叶的图案随意的分布在墙壁和天花板上，电灯凹凸有致，直观地强调了流动性。而第三个房间是个略高的半圆型抽象森林空间，垂吊的各种形状绿色绳子构成抽象森林的意境，而透过窗户，还可以清晰地看到外面的湖景。其他空间以满足人们交流和反省的基本需求。纯白色的房间——无论墙壁上的绳子还是地上的垫子都采用白色。颜色的缺失创建出一个自省空间，供人冥想。

而交流空间则紧挨着瑜伽房，为人们的社交活动提供场所，是一个能让人交谈、休憩以及结交朋友的空间。

青铜镶嵌的窥视孔被特地而随意地安放在瑜伽房墙壁上，既便于人们在做各种瑜伽姿势时注意力集中，同时也满足了人们窥视的天性，通过窥视孔看到一个宁静的世界，把宁静世界融入到自身而达到平静的境界就是通过凝视－感知－排除杂质这个过程开始的。

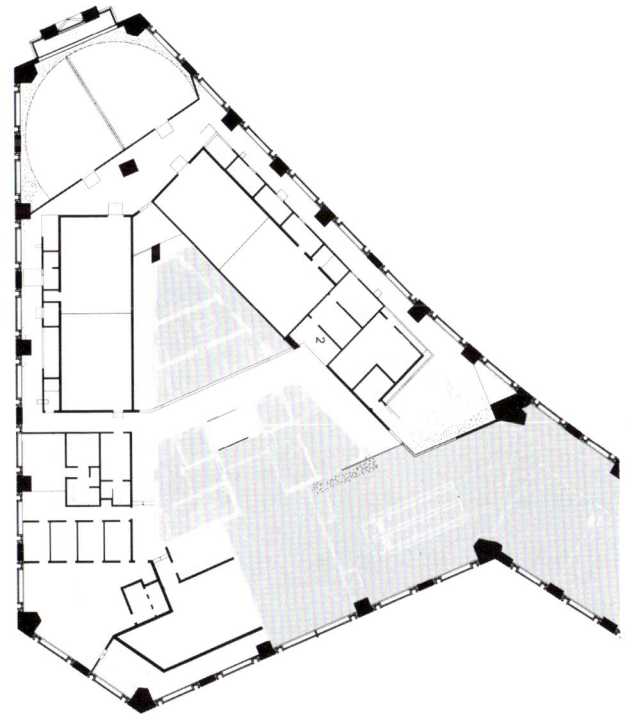

Breaking New Ground

新天地

Project Name: Breaking New Ground **Designer:** Jutta Friedrichs, Ben Houge
Location: Shanghai.china **Photographer:** Jutta Friedrichs, Ben Houge **Time:** 2008

项目名称：新天地 设计师：尤塔·弗里德希，本·雨果 项目地点：中国，上海 摄影：尤塔·弗里德希，本·雨果
完成时间：2008年

100% Design Shanghai celebrates Shanghai International Creative Industry Week with the joint installation "Breaking New Ground," a collaboration between MÜ designer Jutta Friedrichs and sound artist Ben Houge, providing a sneak preview of next year's 100% Design Shanghai show. The 240 sqm joint installation provides a showcase for new local and international design objects, specially selected for the exhibition, which metaphorically break through the ground and float amid shards of ice, as a fresh wind blows in from overhead. This complete sensual experience reflects excitement and anticipation for the gathering wave of design and creativity that is swelling in China.

Designer Jutta Friedrichs and recipient of two Red Dot Design Awards this year, provides the vision behind the MÜ design furniture collection.

为庆祝上海国际创意产业周的开幕，上海百分百设计展在新天地的共享大厅发布了将在明年的展会上展出的产品。会场由设计师尤塔·弗里德希和著名艺术家本·雨果合作设计而成。共享大厅占地240平方米，国际国内的最新设计成果都在这里向公众展示，其展出方式十分特别，是设计师为此次展览所特别挑选的，展品有的像是要冲出地面，有的像是在碎冰上飘浮，像是一阵清风从参观者身边吹过。展览带给人完整的感官体验，它能振奋人的心情，激发人们参于设计的积极性，反映了创造力在中国的蓬勃发展。

设计师尤塔·弗里德希和今年两项红点设计奖的获得者，打造了家具设计作品的未来。

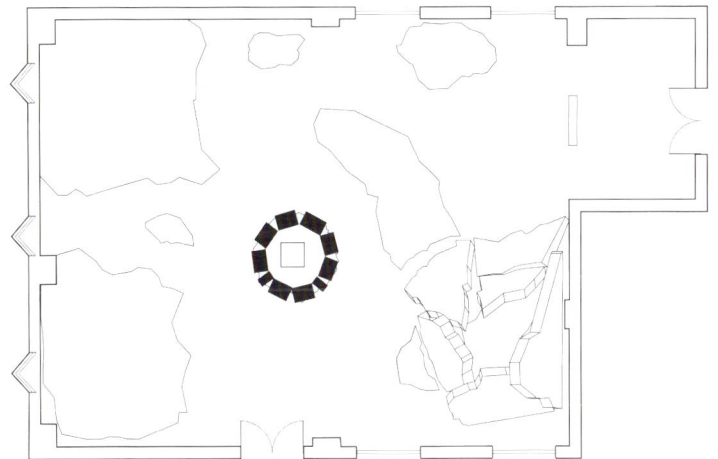

EXIT ONLY

← TICKETS

By studying our vast collections, Smithsonian scientists are h
to piece together the story of how insects and plants co-evol

Botany

Mission: To discover and describe the diversity of plants, explain how they evolved, and understand how humans often and benefit from this diversity.

Collections: One of the ten largest in the world, with 4.7 million specimens representing all major plant groups and tens of thousands of species from all over the world.

Research: Fifty research scientists and affiliated staff focus on the origins, relationships, and classification of plants worldwide, especially the tropics.

Entomology

Mission: To discover, describe, and understand the diversity of insect species, which is numbering up to 30 million, and of which fewer than a million have been identified.

Collections: One of the two largest in the world, with 35 million insect specimens from all over the Earth.

Research: Over 70 research scientists and affiliated staff from the Smithsonian Institution, U.S. Department of Agriculture, and U.S. Department of Defense collaborate to expand our knowledge of insects.

Paleobiology

Mission: To understand the history of life on Earth—how it has evolved, influenced, and responded to recent cataclysmic changes.

Collections: The world's largest fossil collection, with some 40 million fossil plants, animals, and microscopic organisms that record the diversity of life over the last 4.5 billion years.

Research: Twelve staff research scientists and seven associates conduct research on extinction, biodiversity, climate change, and the interactions of life with geological processes, while 15 other staff oversee collections and assist researchers.

Butterflies + Plants

蝴蝶与植物展厅

Project Name: Butterflies + Plants **Designer:** Reich + Petch Design International **Location:** USA
Time: 2008 **Client:** Smithsonian Institution National Museum of Natural History

项目名称：蝴蝶与植物展厅 设计师：雷克+佩奇国际设计 项目地点：美国 摄影：奥索•曾 完成时间：2008年
客户：美国国家自然历史博物馆

In fall of 2007, the National Museum of Natural History opened a new Butterflies+Plants: Partners in Evolution exhibit set within the context of exhibits explaining the co-evolution of butterflies and plants. Reich+Petch Design International was responsible for the conceptualization and design of all exhibits inside the Pavilion and in the surrounding Hall.
The firm worked closely with the museums' scientists and writer to develop the narrative sequence. Reich+Petch Design International also worked closely with the architectural firm responsible for the Pavilion shell structure to ensure efficient visitor flow and seamless integration of displays and environmental support systems.
The concepts were inspired by the colours and forms of the beautiful content and played with scale and imagery to create a delightful experience of fun exploration. The space planning and details were designed to accommodate the museums' exceptionally high numbers of visitors (several millions per year). The exhibit surrounds the visitor with colourful arrays of imagery and specimens to reinforce the connection between evolution and diversity.

2007年秋，美国国家自然历史博物馆开设了名为"蝴蝶与植物"的展厅，旨在探索蝴蝶与植物在进化过程中相互依存的关系。雷克+佩奇国际设计公司受邀负责展厅内部以及相邻大厅的设计与装饰。他们同博物馆的科研人员、作家以及负责展厅外壳结构的建筑公司密切合作，规划空间格局及展品的陈列方式。
他们最终从展品的颜色和形状中获得灵感，运用空间比例及图片营造了一个愉悦而趣味十足的体验之旅。空间的布局及细节设计要考虑博物馆大量的游客，按颜色分类的方式进一步加深了进化过程与物种多样性之间的关联。

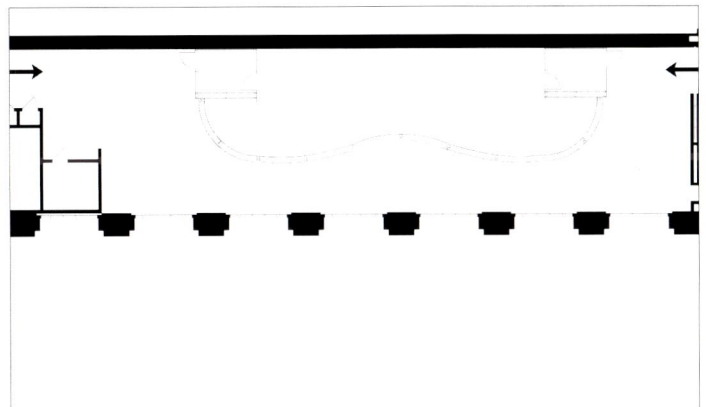

Metamorphosis:

Foes. . .

Transform to . . .

Butterflies and plants interact with each other
—sometimes as friends, sometimes as foes.

Evolving Together

A moth with an 8-inch-long tongue?
And an orchid with a throat to match?
Why go to such lengths?

It appears that over many generations the Giant
Hawk Moth and Madagascar Star Orchid have been
changing in response to each other—a process
called coevolution. Today both partners benefit
from their close tie. The moth gets exclusive access
to food, while the orchid gets a reliable pollinator.

Fuji Xerox Epicenters

富士施乐中心

Project Name: Fuji Xerox Epicenters   **Designer:** Geyer Pty Ltd   **Location:** Sydney, Tokyo, Singapore And Shanghai   **Photographer:** Tyrone Branigan   **Time:** 2007   **Project Size:** 5000-8000sqm

项目名称：富士施乐中心 设计师：赫耶尔设计公司 项目地点：悉尼，东京，新加坡和上海 摄影：泰隆·布莱尼根
完成时间：2007年 面积：5000-8000平方米

Geyer were commissioned by Fuji Xerox, in collaboration with Landor brand and graphic consultants, to create four customer innovation centres across the Asia Pacific to celebrate the heritage brand and reinforce their leadership position in the digital production print market. These "Epicentres" are located in Sydney, Tokyo, Singapore and Shanghai.

The vision of the centres is to offer Fuji Xerox's global customers a complete "digital print" experience – with a focus on business solutions not just "hardware"/equipment.

Visitors are provided with a personalized journey depending upon the reason for their visit and the level of existing knowledge of the Fuji Xerox products and services. A series of "experiential zones" are personalized for each customer, ensuring relevance of brand communication through to sales conversion and customer retention experiences.

From arrival the customer's experience is extraordinary. An iconic sphere, projecting welcome messages to the customer, denotes the entry point. Visitors are greeted by a roaming attendee, unrestrained by the confines of a reception desk. The traditional reception zone has been replaced by a dynamic Fuji Xerox brand "Immersion Zone" where visitors are subliminally immersed into the total brand experience through sound and video content.

赫耶尔受富士施乐公司的委托，与兰德尔品牌和造型顾问合作，在亚太地区建立四家客户创新中心，为古老的富士品牌宣传造势，强化其在数字印刷市场的领导地位。这些创新中心位于悉尼、东京、新加坡和上海。

中心的作用是为全世界富士施乐公司的消费者提供全套的数字印刷体验——着重展示公司的商务运作方式而不仅仅是其设备和硬件。

中心根据参观者不同的到访目的和对富士施乐公司产品和服务不同程度的了解，为他们量身打造个性化的参观之旅。丰富多彩的"体验区"能满足不同客户的需求，强调了品牌传播在销售转换和客户管理方面的重要性。

这里带给顾客的体验是无与伦比的。中心的入口处为参观者传递着欢迎的信息，带他们进入奇妙的世界。参观者由工作人员引导，没有接待台的约束。与传统的接待台不同，这里的大屏幕播放着充满活力的富士施乐品牌介绍，参观者的全部身心都融入整个品牌体验之中。

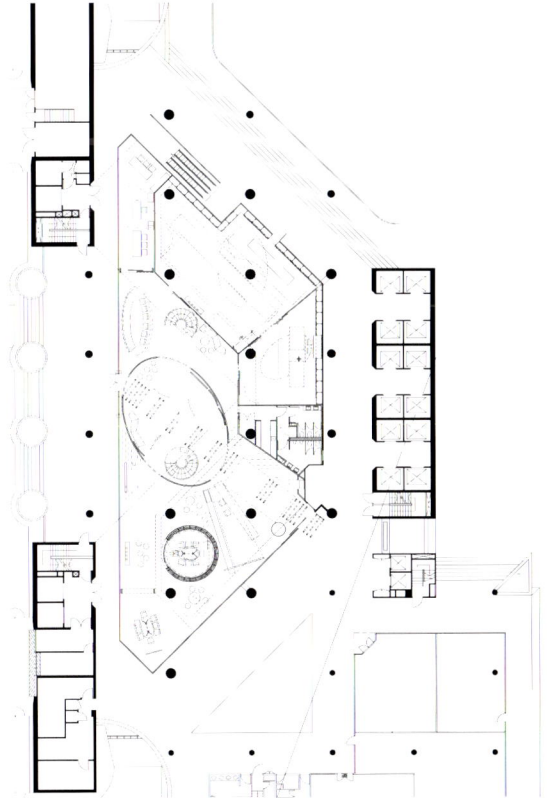

Stylecraft Showroom

风尚家居展示厅

Project Name: Stylecraft Showroom   **Designer:**Geyer Pty Ltd   **Location:** Sydney, Australia
Photographer: Andrew Sheargold   **Time:** Sydney 2007   **Area:** 500-600 sqm

项目名称：风尚家居展示厅 设计师：赫耶尔公司 项目地点：悉尼，澳大利亚 摄影：安德鲁·谢尔格德
完成时间：悉尼，2007年；新加坡，2008年 面积：500-600平方米

Stylecraft is a leading provider of high quality commercial, hospitality and residential furniture to the architectural and design community, representing original contemporary design from Arper, Akaba, Bonestil, Cornwell Hecker, Dynamobel, ESO, Parri, Stua and Verzelloni.

Stylecraft wanted to create a dynamic and innovative space that draws customers to the showroom — and encourages them to return.

Geyer's design solution embodies Stylecraft's brand attributes – energetic, confident, kinetic and fun.

The space has been designed to evolve and change with the business, and enables a diversity of customer experiences. A highly flexible environment has been established that allows the space to constantly evolve and change to support a range of local and international brands and new product lines.

Two key environments have been created, the showroom and sales work area. While there are no walls separating these areas, the use of different finishes and lighting dramatically affects the mood and creates a clear distinction between the spaces.

In a move away from the traditional "chair wall", mobile chair towers are suspended from the ceiling and slide effortlessly through the showroom. This gives the chairs (Stylecraft's core product offer) prominence within the space. The mobile towers also allow the space to be quickly and easily reconfigured to cater for Stylecraft product launches and industry events.

风尚家居有着高品质的家具，适合商用、酒店和住宅，是建筑设计界的首选。其主要合作伙伴包括阿布尔，亚喀巴，博塞尔，康韦尔·海克，代那莫贝尔，ESO，帕里，斯图尔和瓦泽洛尼。

风尚家居希望展厅充满活力和创意，吸引顾客前来参观，并让他们愿意再次光临。

赫耶尔的设计体现了风尚家居的品牌特性——充满活力、自信、动感和乐趣。

展厅的设计随着展示主题的不同而发展变化，给顾客带来不同的体验。展厅十分灵活，本地品牌、国际品牌和新产品的展示区各不相同。

展厅内有两个主要的空间，展示区和销售区。二者之间并没有用墙作分割，而是采用不同的装饰与灯光形成二者之间的明显差别。

与传统的"椅子墙"不同，设计师将椅子放在可以移动的梯格上，悬挂在天花板上。以这样的梯格贯穿整个展厅。风尚家居的核心产品——椅子在展厅内十分突出。这些椅子还能快速而简单地重塑空间，以配合风尚家居新产品的推出和行业活动。

Haloid 3

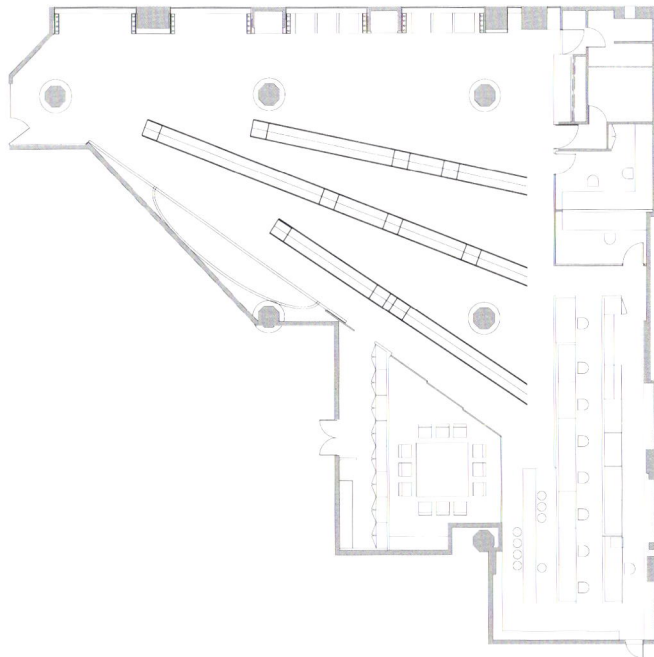

Bernhardt Design Showroom

伯尔尼设计公司展厅

Project Name: Bernhardt Design Showroom **Designer:** Lauren Rottet, Kelie Mayfield, Christopher Evans **Location:** Chicago, USA **Photographer:** Eric Laignel **Time:** 2008

项目名称：伯尔尼设计公司展厅 设计师：劳伦·罗泰特 凯利·梅菲尔德 克里斯托弗·埃文斯 项目地点：美国，芝加哥 摄影：埃瑞克·莱格尼尔 完成时间：2008年

The Bernhardt Design Showroom recently moved locations within the Chicago Merchandise Mart in May 2008 to the opposite side of the building, down the hall from the previous showroom. The move created a challenge for the design team since the new space no longer had the wonderful views of the river and the showroom decreased considerably in size to only 7,000 square feet. This move caused a significant challenge to adequately display the client's furniture product. The design team's solution was to create a space that would emulate a museum setting for the furniture to be displayed as art.

The biggest challenge the design team was charged with was to create a small space that feels much larger than its actual size. The design team utilized light as the medium for solving this challenge and the source for extending the perception of space. Light was used so that solid planes never touch floors, walls or ceilings and always have the appearance of floating. Light paneled walls were also used as accents to reflect light so as to create the appearance that natural light leads one into the space beyond – as if the front door was opening and allowing daylight to enter the space.

The space also had no interesting views to the outside world so the design team decided to create a world inside the showroom and focus the attention of the viewer on the art – the furniture. The design team's solution was to create an ephemeral wall which would also double as a backdrop for the client's chair display. The wall is comprised of metal painted white with a custom designed perforated pattern. Fabric scrim was stretched behind the panels to wash the light fixtures and create a consistent source of light. The floor pad aligning the paneled wall is made up of shiny white epoxy resin to help reinforce an idea of a plinth, but yet accessible and inviting for one to experience the chairs and become part of the art scene and experience.

The finishes used in the space also played a significant part in the overall design. The materials used in the architecture and furniture are all "visually quiet" to not interfere with the lines and detailing of the furniture. The space is void of colour and the fabric and carpet used hints at the idea of wool or cotton before it is dyed and coloured.

伯尔尼设计公司展厅从原有的位置搬迁到大厦的另一侧，这一举措为设计团队带来了许多挑战：新空间的窗外再也不见美丽的河边景致，面积也减少到仅为650平方米；如何在有限的空间内展示家具产品。为此设计团队构思了巧妙的解决办法——打造一个博物馆般的背景，让每件家具都成为艺术品。

首先，为解决如何在视觉上增添空间开阔感的问题，设计团队充分运用光线元素。隔断板结构与地面、墙壁及天花分离开来，营造悬浮感；光嵌板墙壁用于反射光线，使得每一道光线仿佛是从门口照射进来一般。

其次，为弥补室外景致的缺失，设计团队将目光全部吸引到家具展品上，营造了一个布满艺术品的精美室内世界。临时墙壁作为展品背景，展示着各种椅子。白色油漆粉饰的金属结构墙壁带有专门设计的孔状图案；平纹棉麻织品在墙板后面延伸，形成了连续的光源；地板垫及墙面板材全部由白色抛光环氧树脂打造，成为了展览的一部分。

最后，装饰元素同样扮演着重要角色。建筑材质及家具装饰的选择以"淡雅平和"为理念，目的是不抢占展品的风头。白色构成了主要基调，即便是织品装饰及棉毛地毯也未经任何染色处理。

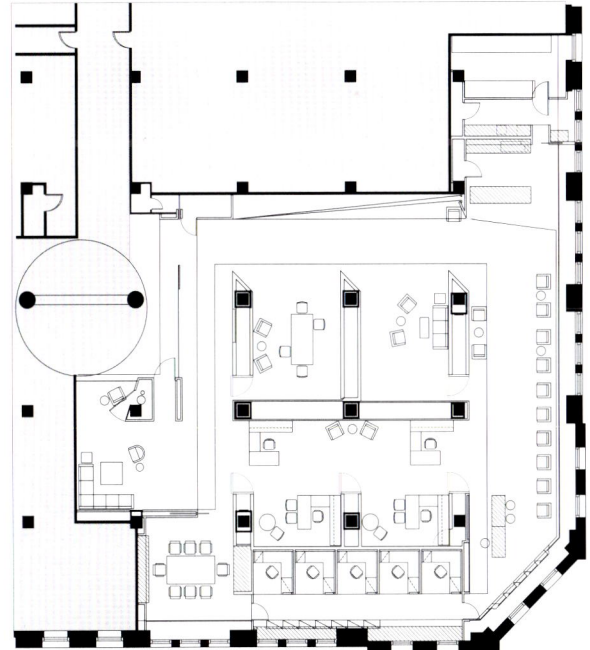

BRANCH in Changchun

长春藤

Project Name: BRANCH in Changchun   **Designer:** Keiichiro SAKO, Yoshimasa TSUTSUMI/SAKO Architects   **Location:** Changchun, China   **Photographer:** Misae HIROMATSU (BEIJING NDC STUDIO, INC.)   **Area:** 280 sqm

项目名称：长春藤 设计师：SAKO建筑师事务所 项目地点：中国，长春 摄影：北京和创图文制作有限公司 面积：280平方米

One corner of the Changchun Library of Jilin Province is a culture exchange centre founded by International Exchange Foundation that devoted to introduction of Japanese culture to Chinese people. In order to create a relax space with easy access to people for party or personal ease, where they can find their respectively satisfactory and pleasant things, the designers designed it by centring on trees. Any designs in the room are characteristic of a forest, attracting and charming, with people seating or leaning on the branches and hiding between branches. It is a space that arouses people's imagination.

The complicated design at first sight is factually based on human structure, including davenport, sofa and desk, all much convenient for people. The material adopted is the high-dense plate and is assembled at the site after rough preparation in Beijing. It has fully reflected the highly efficient construction management and operation. The floor, wall and ceiling are generally white and the original pillar is covered with stainless steel mirror to weaken its existence. The design, green in general, gives relaxing touch to the whole space and the color gradation, together with the LED lamp, has made much of its vitality. Besides, sphere-shaped LED lamps are also placed everywhere in the room to convey the morning dews impearled on leaves.

It is a place full of vines and tendrils that produces diversified achievement and it is an organic and complimentary unit formed by inter-twisted vines. The division and unity of the tree have just depicted the exchange and growth of the relationship between China and Japan.

长春藤——中日文化交流中心坐落在长春图书馆的一角。为营造轻松愉悦的氛围，设计师特别在房间中引入了小树，这样人们便可以靠在树干上或藏在树丛中尽情地交流思想。

设计元素看似复杂，其实完全根据人体结构打造，比如沙发、桌子等非常便于使用。地面、墙壁以及天花板全部饰以白色，原有的梁柱结构采用不锈钢镜面覆盖，以与整体氛围相符。绿色元素的引进更是增添了恬淡舒适的气息，半圆形的LED灯随意地摆放着。总之，这是一个藤条缠绕的空间，预示着中日友谊的不断发展。

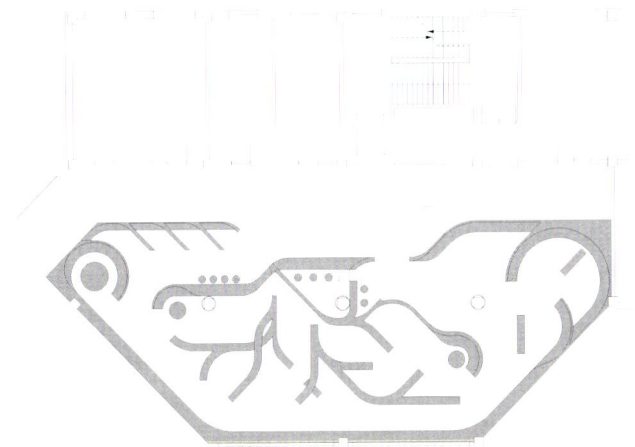

Free Game

FlatFlat in Harajuku

原宿平平游戏体验店

Project Name: FlatFlat in Harajuku   **Designer:** Keiichiro SAKO, Yuichiro IMAFUKU / SAKO Architects
Location: Tokyo, Japan   **Photographer:** Chikao TODOROKI / Sasaki Studio Inc.   **Time:** 2008
Area: 328 sqm   **Client:** NHN Japan Corporation

项目名称：原宿平平游戏体验店 设计师：SAKO建筑师事务所 项目地点：日本，东京 摄影：佐佐木工作室 完成时间：2008年
面积：328平方米 客户：日本NHN集团

FlatFlat in Harajuku is the store where the visitors are actually able to experience Hangame, an online games portal, and Hange.jp, a games portal for mobile phones, developed by NHN Japan Corporation. Located at the centre of Harajuku, the store occupies a narrow space of 3.5 metres wide by 45 metres long. The concept of the store is the future park and the designers attempted to combine virtual element with real space opens to anyone. It is not only the space where 'organic principle' and 'inorganic principle' stay together, but it represents the modern society consisting of real and virtual environments.

"Organic principle" consists of curved lines and has a space characteristic of a cave that people can discover depending on each purpose and feeling while the different scenes unfold continually. They attempted to create a space that people feel like snuggling up to the organic form that curves based on human body dimensions. Visitors do not hesitate to enter the store because of the sense of closeness. On the other hand, "Inorganic principle" consists of white wall surface fixtures, neon tubes of ceiling illumination and mortar floor creates virtual character. They also used the lines which let neon tubes and the prevention of crack seam of mortar offset the forms of the wall surface fixtures. They stimulate curiosity of the visitors by synergy with the forms of the fixtures and lead them to the inner part of the narrow space.

平平游戏体验店坐落在原宿，为顾客打造了一个真实的网络游戏体验平台。设计师以"未来公园"为主要理念，将虚幻的元素和现实的空间完美融合。

如同洞穴一般的空间以弯曲的线条为基本元素，人们可以根据其不同的用途分辨并使用，真切地感受到它们的存在。模仿人体结构的独特设计风格更是让顾客感到真实，吸引着他们走进来。另一方面，墙壁上的装饰、屋顶上的霓虹灯条不断的闪烁着，营造了虚幻空间感，对于顾客来说，这无疑更能激起他们的好奇心，一探究竟。

Communication

Free Game

ふらっときてね。

bb flat flat
hangame square

homeculture

文化屋

Project Name: homeculture   **Designer:** Franken Architekten   **Location:** Berlin, Germany
Photographer: Franken Architekten   **Time:** 2005

项目名称：文化屋 设计师：弗兰肯设计事务所 项目地点：德国，柏林 摄影：弗兰肯设计事务所 完成时间：2005年

The designers would like to provide a spatial context for the entrepreneurial vision of the client – visible, tangible, and palpable to all the senses. They are never just functional, but rather communicate and transport messages effectively and intuitively.

So they present the designers a showroom like a high-end fashion store for a selection of tiles and spa assecoires.

The store provides an ideal presentation platform for Raab Karcher and its four joint-venture partners from the premium tile and bathroom fittings sector.

The designer used animated lines of vision in order to generate virtual space.

As the onlooker's gaze wanders along the front was taken as the basis for generating the interior's geometry by means of computer simulation. The fact that it is shallow in depth means that for passers-by the store functions as an elongated shop window.

Visual-kinetic installation is another feature of this project.

The Liquid Wall appears like a vertical water wall and serves as an eye catcher in the Ku'Damm facade. In a computer simulation, the passer-by's gaze causes an acrylic glass pane to vibrate.

The frozen vibration acts like a distorting lens and makes the illuminated back wall oscillate as you wander past it

客户希望空间中所有的感官享受有形化、实际化，并且触手可及。它们不仅具有本身的功能和作用，还起到有效地交流和传递信息的作用。

于是，设计师打造了这间文化屋，它就像一间高档装饰材料和水疗设施的展示厅。

这里的瓷砖和浴室配件全部采用拉布卡切和其他四家合资品牌的产品。

设计师运用动感的线条营造出视觉上的虚幻空间。

电脑仿生手法做出顾客在店前流连的影像，并以此作为店内几何效果的基础。这一效果离窗口较近，它将过路的人们记录下来，是橱窗的延伸。

视觉动力学装置是项目的另一个特点。

水幕墙是一堵垂直的水墙，它面向库丹大街，因此十分引人注目。利用电脑仿生手法，过路人的目光会使亚克力玻璃面板产生微微的震动。

震动的面板就像扭曲的镜子，当你经过背光墙时，连它也会摇摆起来。

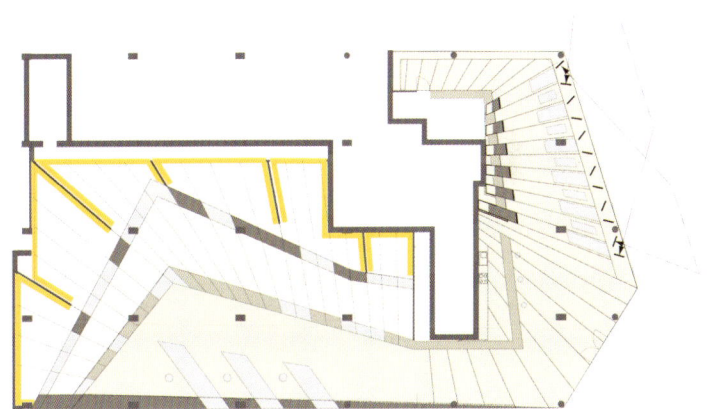

GRAFTWORLD-Exhibition in the Aedes Gallery

Graft作品展厅

Project Name: GRAFTWORLD-Exhibition in the Aedes Gallery   **Designer:** Graftlab
Location: Berlin, Germany   **Photographer:** Jan Bitter   **Time:** 2007

项目名称：Graft作品展厅 设计师：Graftlab建筑设计事务所 项目地点：德国，柏林 摄影：简·比特 完成时间：2007年

In addition to GRAFT's the widely published projects, like the prize-winning Hotel Q in Berlin, Restaurants Fix and Stack in Las Vegas and several international houses, the exhibition will focus on the new and unpublished work of GRAFT, including numerous highrise buildings, special hotel designs and private residences, research installations as well as public projects.

GRAFTWORLD refers to GRAFT's specific method of work process, which fuses syntactic, semantic and phenomenological aspects into architecture. GRAFT uses narrative elements of cinema, exploiting space — time interconnections through storyboarding with progressive technical research of long term collaborations with various companies.

The exhibit showcases this methodology in an accessible, usable exhibition sculpture, an interactive lounge displaying projects of all three office locations. Classical distinctive architecture elements like floor, wall and ceiling as well as the classical disciplines of additive plasticism and substractive sculpture will be fused into a lounge-display-hybrid. GRAFTWORLD is thought to be inclusive and rather crosses and blurs boundaries than accepting their exclusive nature.

GRAFTWORLD proposes architecture that can evolve to be sculpture, be furniture, be a product, be an urban scenario, be landscape at the same time. Design, methodology and technology are not classically "displayed", but are integrated in the GRAFTWORLD .

Graft曾创作无数众所周知的作品并被刊登在各种媒介上，包括Q酒店、菲克斯餐厅灯等。当然他们还有很多作品未被发表，这一展厅就专门用来展示这些创作。Graft天地主要是指他们独特的创作方式，将句法语义等语言特色融合到建筑中。为展示自身的特色，他们选择了通俗易通的模式——将坐落在不同城市的三个机构的作品汇集在一起，并运用地面、墙壁以及天花板等传统的建筑元素，将它们混合在一起完美演绎，如同一件雕塑、一件家具、一个产品或是一处景观。设计、方法以及技术融合在一起，不愧为大胆的方式！

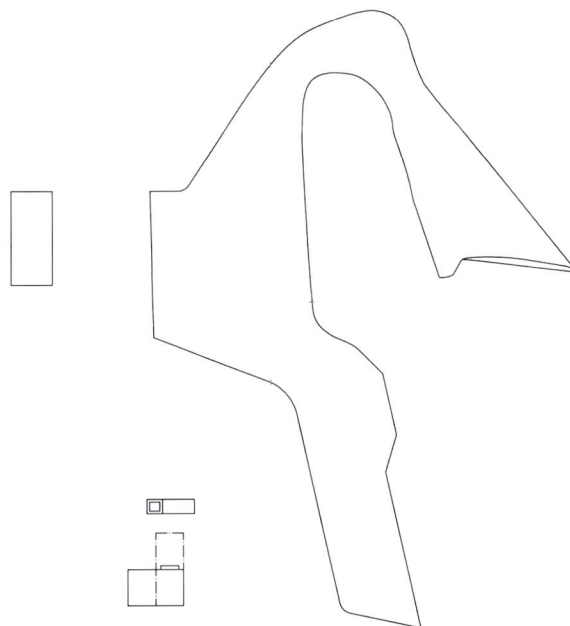

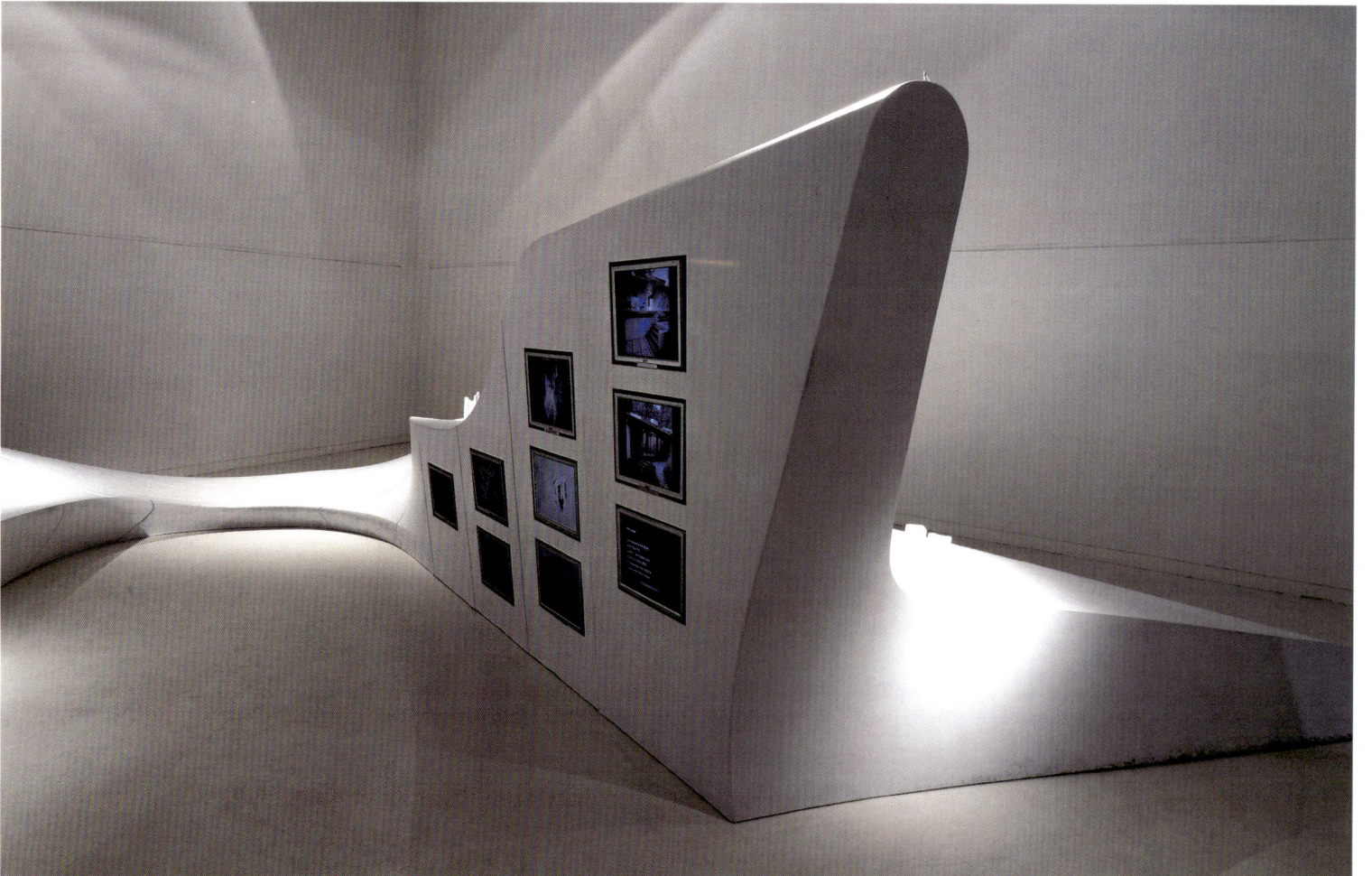

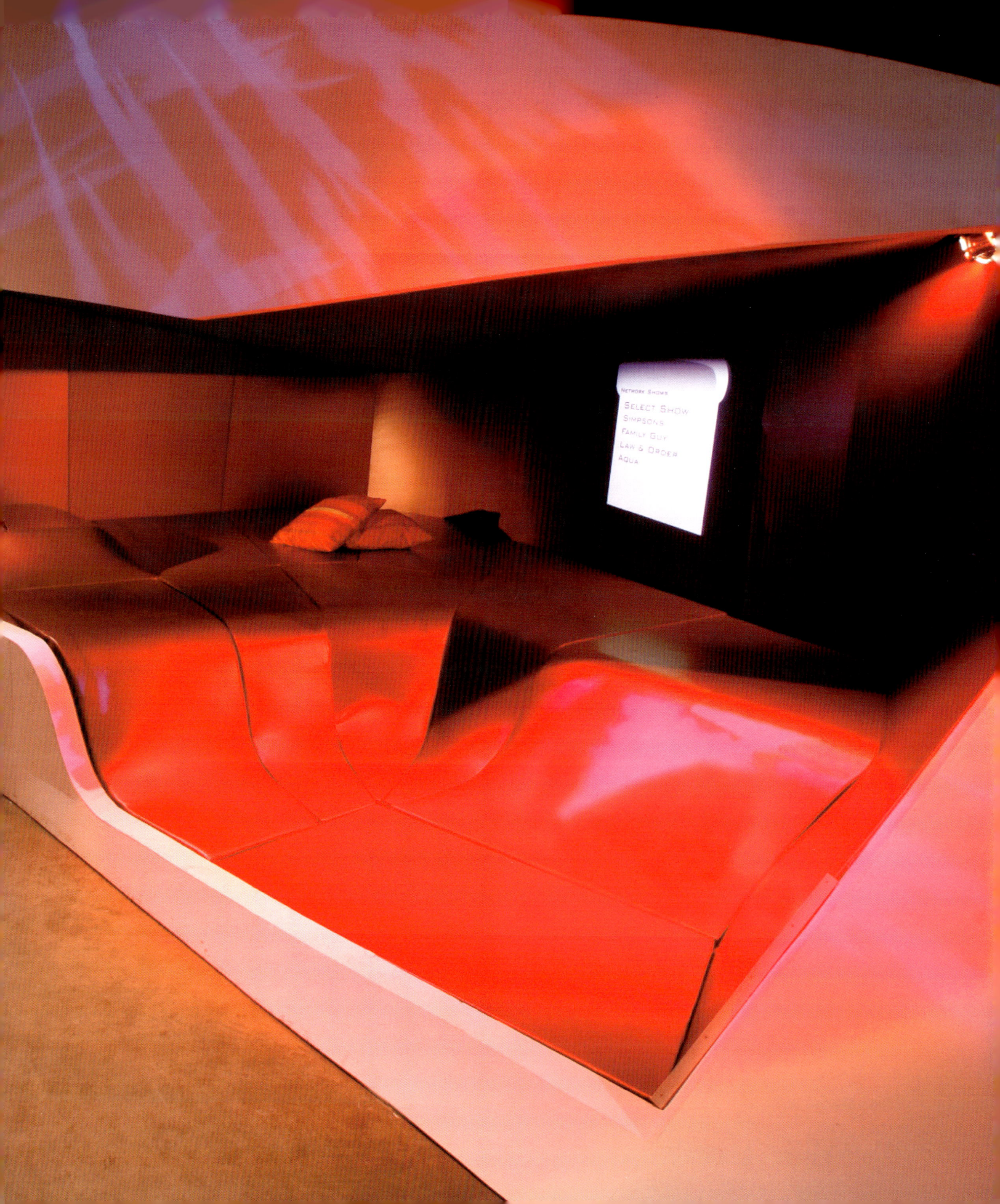

NETWORK SHOWS

SELECT SHOW
SIMPSONS
FAMILY GUY
LAW & ORDER
AQUA

Moonraker

蒙莱克尔

Project **Name:** Moonraker   **Designer:** Graftlab   **Location:** Burbank USA   **Time:** 2006

项目名称：蒙莱克尔 设计师：Graft建筑设计事务所 项目地点：美国，伯班克 完成时间：2006年

Graftlab will design and build these worlds as stages of visual communication. Like a set design, these interior worlds will be erected within a hangar or loft space. They will become accessible statements, places of experimentation and inspiration.

Concept

To provide a tangible, engaging experience representative of future Volkswagen customer needs and desires, an advanced look at themes will be of growing importance in the coming decade. The Life Settings will be a powerful tool to support and inspire a multitude of projects. Four futuristic scenarios illustrate a broad spectrum of user types, interests, competencies and aesthetics.

Aspirational architecture will showcase forecast technologies and concepts in a progressive sequence, preceded by a walk through the current consumer kit presentation and finishing with an automobile sized viewing "drive way". The installation can be modified to suit a variety of foreseeable scenarios and uses.

这一项目目标即为打造如同舞台一般的视觉交流空间，设计师运用布景设计原则，将这些不同的"世界"安置在阁楼空间内，让其成为体验与灵感之所。

理念

为满足大众汽车潜在客户的需求，设计师决定打造一个触感十足而又格外引人注目的空间环境。四个未来风格的"世界"以特有的顺序展示着先进的工艺技术及设计理念。一条小路紧接着蜿蜒前行，穿越到其他展区。尤为值得提到的一点是，这一空间可以根据需要随时改变，以满足不同的需求。

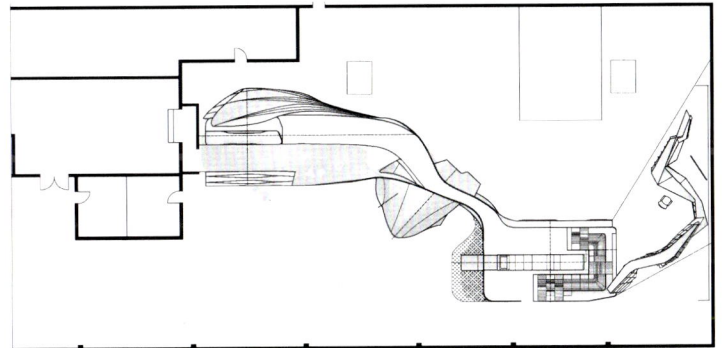

Internorm Flagshipstore

Internorm办公展厅

Project Name: Internorm Flagshipstore **Designer:** ISA STEIN Studio **Location:** Pasching, Austria
Photographer: TEAM M/ISA STEIN **Time:** 2007

项目名称：Internorm办公展厅 设计师：I艾萨·斯坦恩设计工作室 项目地点：奥地利，帕兴
摄影：TEAM M/艾萨·斯坦恩设计工作室 完成时间：2007年

The Frame

Unique in its design and grace, the "archi-FRAME" from Internorm is setting new standards in the architecture of a CI-suitable Flagship-Store. The frame is an integral part of Internorm's "corporate identity" – the company identity. One can look into the showroom via the building-high "archi-FRAME". Thus the showroom itself is presented through a larger frame as is the production of windows and the shell which would otherwise play a secondary role.

The "archi-FRAME" is made of coloured glass and thus establishes a relationship to the main material of the windows. The glass is used here as a constructional element and not as a filling as is normal. During the night the lettering INTERNORM is illuminated on the frame which is likewise lit up.

Insights, Views, Openings

In the interior, the views and insights were of particular importance to the Architekten Team. In particular a visual contact was to be established to the showroom from the first floor. Thus one can look from the kitchen or the meeting room through a belt of windows into the sales room.

Apart from the frame, the core of the company, the use of images was also important. The aim was to liven up the room via a wall projection. Thus various subjects from Messrs. INTERNORM find a place in the showroom. The images are also to be found in a niche with a seating arrangement and in the conference room.

框架

独特的框架结构作为Internorm企业标识的一部分，设计精良而外观优美。透过框架远远望去，展厅本身更像是一个展品——摆放在框架内的"主角"展品。

框架由彩色玻璃打造，与窗户所用材质形成独特联系。在这里玻璃更像是一种建筑元素，夜晚"INTERNORM"的标识在框架内闪闪发亮，如同被点燃一般。

视野、景观、窗户

室内设计中，视野及景观起到尤为重要的作用，尤其是展厅与外界的视觉联系。为此，设计师连续打造了一排窗户，这样视线便可穿过厨房或者会议室一直到达展品展售中心。

除其特有的框架结构，图片的运用也同样成为公司理念的一部分。设计师巧妙地选用了幕布墙，不仅可以展示公司的产品，同样增添了空间的动感。此外，在座区或是会议室同样可以看到各种图片。

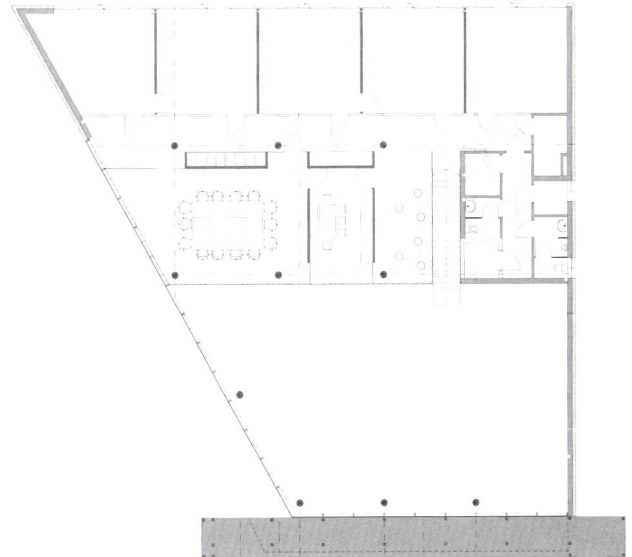

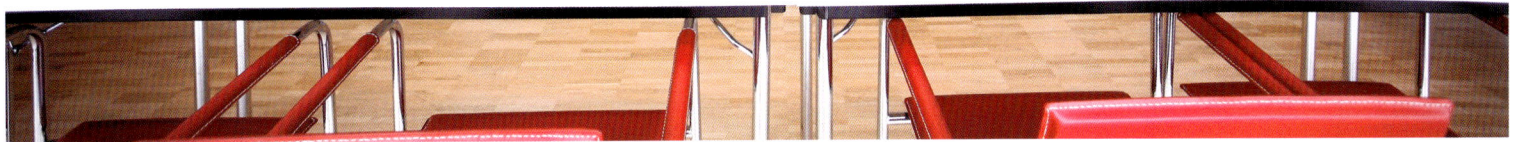

Heaven

悦天展示

Project Name: Heaven   **Designer:** Capella Garcia Arquitectura   **Location:** Barcelona Spain
Photographer: Rafael Vargas   **Time:** 2008

项目名称：悦天展示 设计师：卡帕拉·加西亚建筑事务所 项目地点：西班牙，巴塞罗那 摄影：拉斐尔·瓦格斯
完成时间：2008年

"Heaven" invites the visitor to travel along a tunnel revealing the material's varied possibilities. There are two openings in the facade: one of them, which comes forward to receive you, is the entrance, and the other, which seems to retreat from you, is the exit. The interior lighting keeps changing colour and shade, making the plastic membrane into something living. As you enter you go through a large dome where an oscillating lamp hangs like a tonsil. In one place, the glossy-finish plastic material looks like a painted wall and reflects the light; in another, the same material, but this time with a translucent finish and illuminated from behind, turns the walls into a great lamp that keeps taking on different colours, through a system of electronically-programmed LEDs. When you reach the end, the black walls make the space disappear, and in the darkness the moving image of a fire emerges and vanishes behind the microperforated surface of the material, deceiving the eye and producing a hyper-realistic effect. Opposite the image, in the central zone, one whole wall is filled with a projected video of the sea-bed in constant motion, giving the sensation of being submerged under water. Continuing towards the exit you come to an oval window where the material becomes semitransparent, and through it can be seen a curtain of autumn leaves in silhouette, in constant movement thanks to a fan.

It has been like a journey through an organic space, a series of parabolic, warped shapes that would be impossible to make in any other material. When it is taken down, the entire installation will fit into a small, light case. This plastic material avoids the need to build walls and ceilings with plaster and paint, requires practically no maintenance and is very quick to install. Its lightness means that it is easy to transport and it is totally demountable and recyclable. All in all, it is a good option for ecodesign and sustainability.

悦天展厅由一个管状隧道结构组成，设有两个开口，其中略微前倾的是入口，而稍显后移的为出口。室内灯光不断变换着色彩，使得塑料薄膜材质看上去如同在"跳跃"一般。高大的穹顶空间内，仿若扁桃体形状的灯具悬挂在上。在一段空间内，经过抛光处理的塑料结构看起来更像是一面喷漆墙壁，反射着闪烁的光线；在另一端空间内，同样的材质经过透明装饰以及背光照明处理则变成了一个巨大的灯饰，散发出不同色彩的光线。当到达末端的时候，黑色的墙壁呈现，整个空间在刹那间消失，移动的火光在带孔材质后面若隐若现，营造了一个超现实的氛围。火光的对面，正面墙壁投放着海底世界的画面，让人在现实与幻境之间穿越。继续前行，便可达到一扇椭圆形的大窗内，材质变得半透明，秋叶飞舞的画面呈现在眼前。

必须提到的一点便是这一特殊材质——塑料，不仅省去了墙壁及天花结构的打造，同时不需要精心保养，而且能够在短时间内安装完成。轻巧的质量则使得其特别方便运输，尤为重要的是符合环保标准。

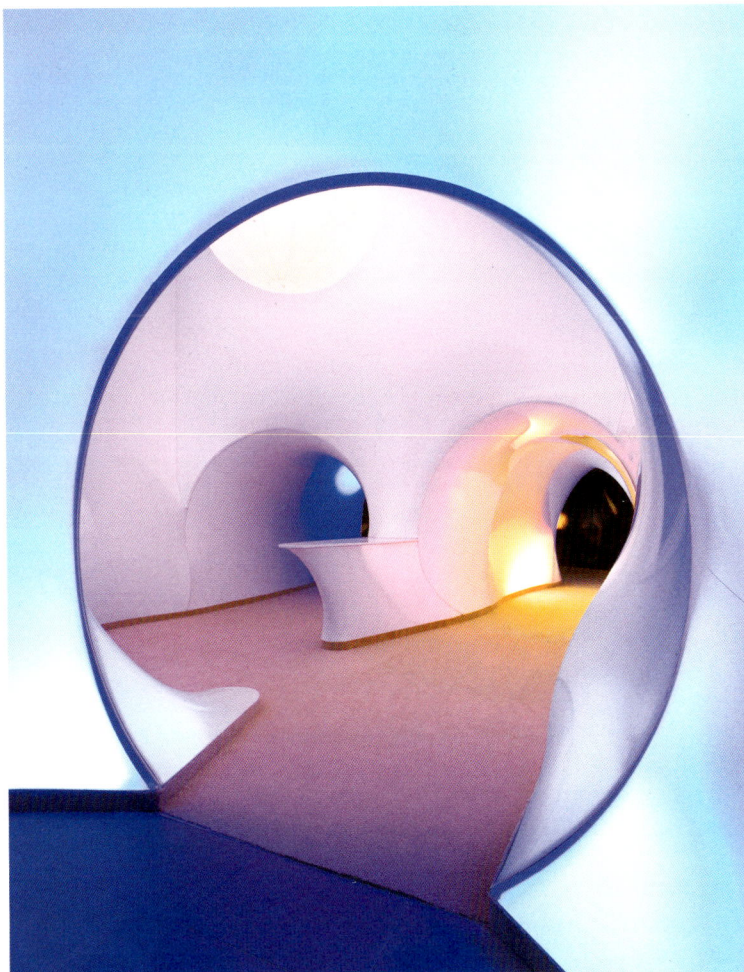

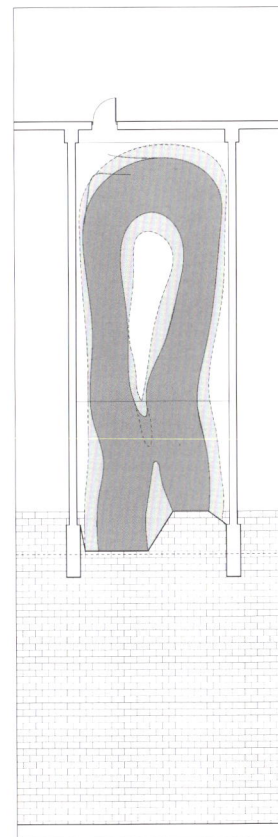

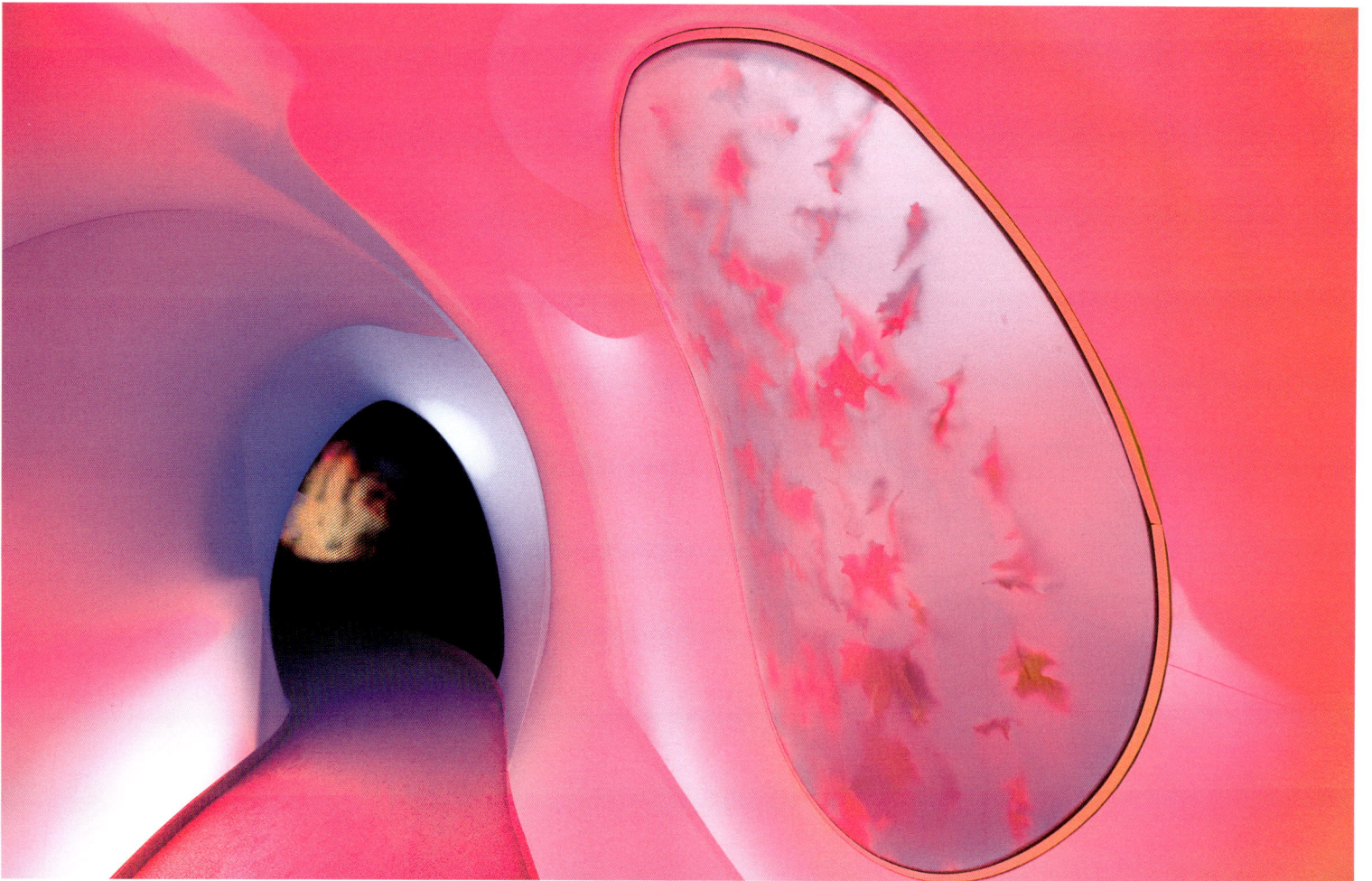

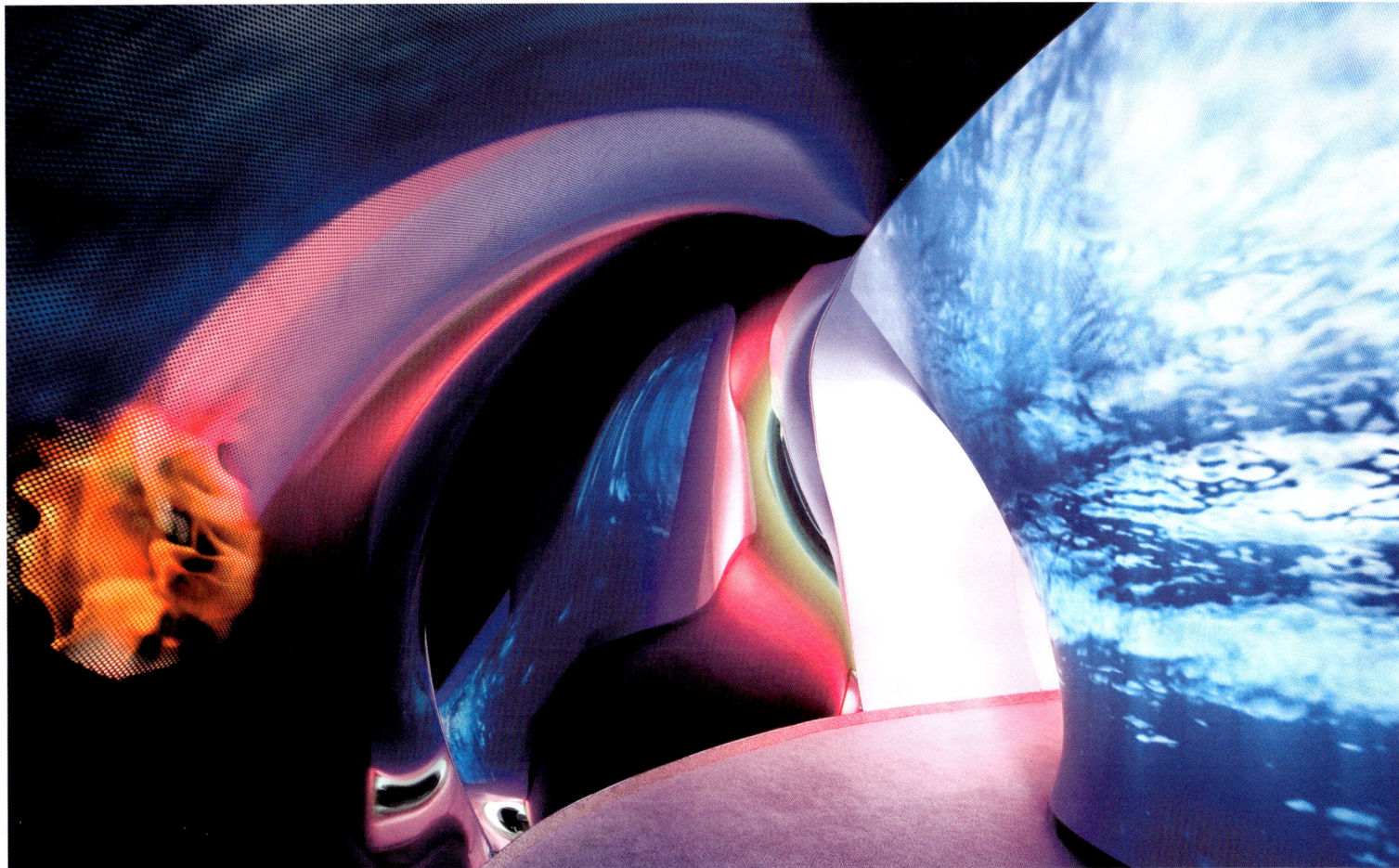

Tesla Store

特斯拉汽车专卖店

Project Name: Tesla Store   **Designer:** CCS Arhictecture   **Location:** Los Angeles, USA
Photographer: Eric Laignel   **Time:** 2008

项目名称：特斯拉汽车专卖店 设计师：CCS建筑公司 项目地点：美国，洛杉矶 摄影：埃瑞克·莱格尼尔 完成时间：2008年

Tesla Motors, maker of the world's most advanced electric car, has opened its flagship Los Angeles location, featuring architecture by CCS Architecture, in May of 2008. Like Tesla's electric sports cars, the store design emphasises style, efficiency and performance.
Located on the corner of Santa Monica and Sepulveda boulevards, the $2 million dealership sprawls 10,000 square feet and features poured concrete floors, an exposed ceiling of ductwork and beams, mirrors and potted plants.
CCS has merged to create an efficient, gallery-like setting for showing and servicing the cars. "The cars are beautiful, so the showroom is set up to be a boutique experience, where the intended rawness of the space counterpoints the sleek refinement of the cars. It's like sculpture in a gallery," says Smith, a car dealerships. The design unifies the showroom and service areas, making the service process fully transparent to guests as well as passersby on Santa Monica Blvd. With an electric car, there is no noxious smoke, no puddle of oil or gasoline, and no loud engine. The Tesla mechanic, like a chef in an open kitchen, has nothing to hide.

作为世界最先进的电车制造商，特斯拉汽车于2008年5月在洛杉矶开设了旗舰店。同其以往的跑车专营店一样，店内设计注重风格、效能及实用性。
专营店坐落在圣莫妮卡和塞普尔韦达大街拐角处的大厦内，耗资200万美元，占地约930平方米。灌浇混凝土地面、天花板上管道及屋梁裸露在外，镜子以及盆装植物极尽简约。
设计的宗旨是打造一个高效的、美术馆风格背景，用于展示商品。"这些车太吸引人了，所以展厅一定要营造一种精品店的体验，用空间的粗糙突出车的精致，在合适不过了。"其中一位设计师这样解释。展厅同服务区设在一起，更让顾客和路人清楚地看到整个服务过程。

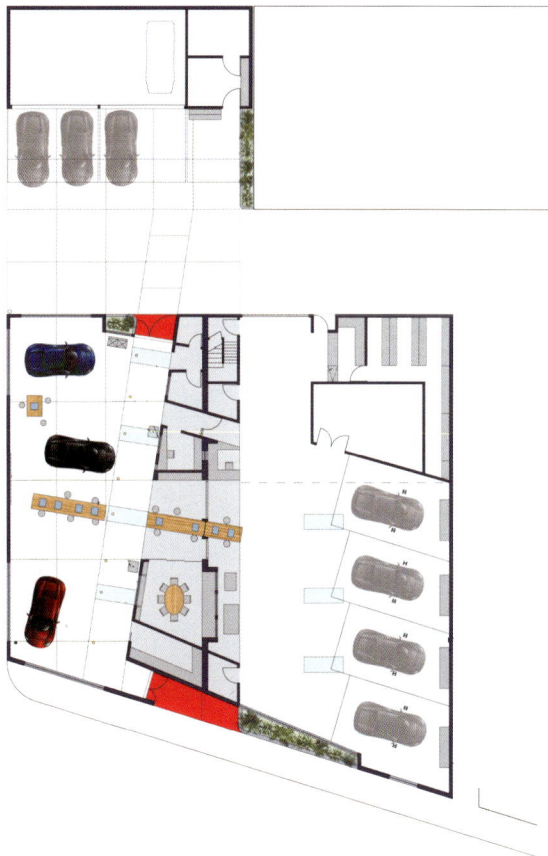

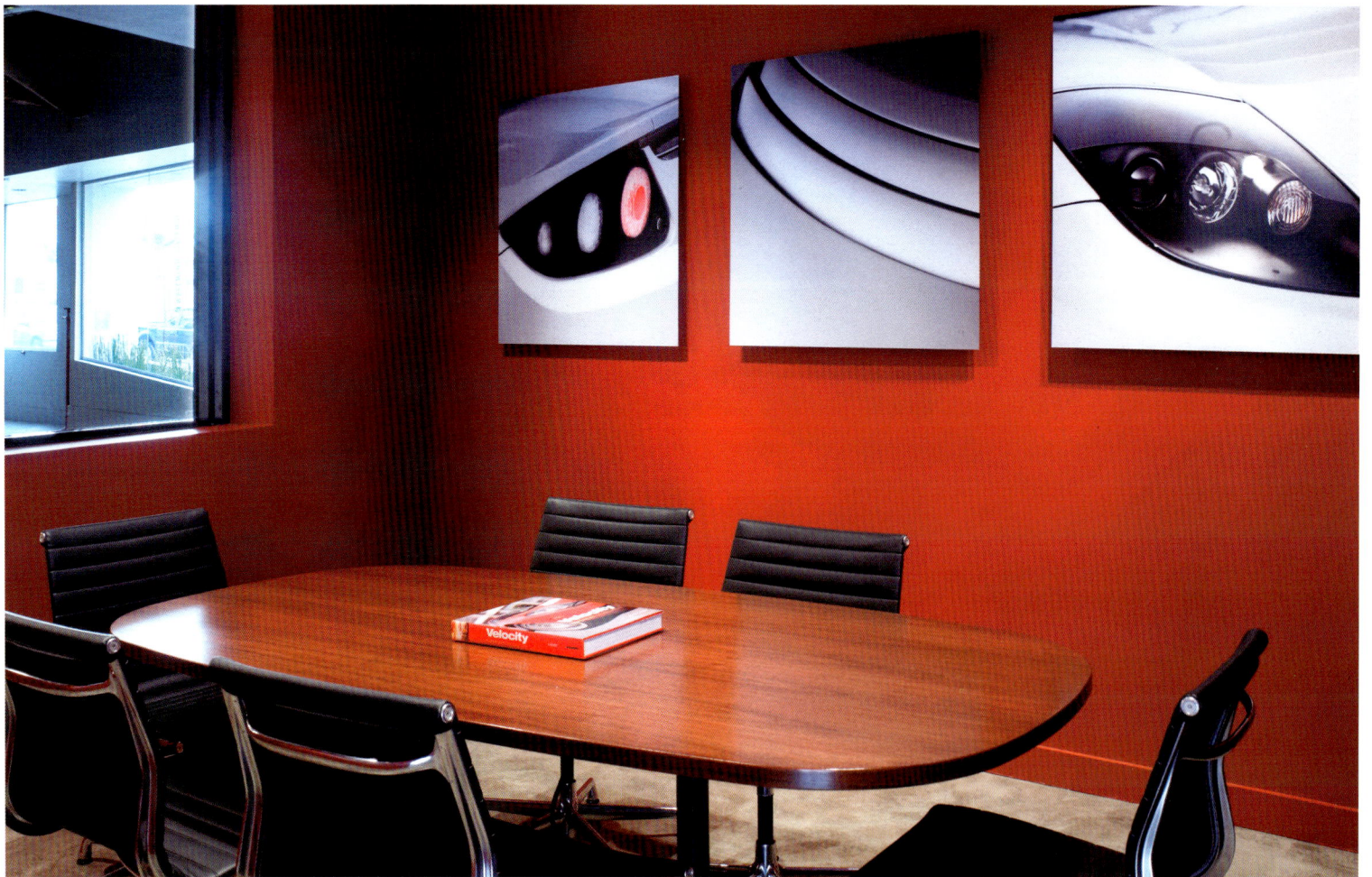

ProNature

天然坊

Project Name: ProNature　**Designer:** Joey Ho / Joey Ho Design Ltd　**Location:** Elements, Kowloon Station, Hong Kong　**Photographer:** Ray Lau　**Time:** 2007

项目名称：天然坊　设计师：何宗宪，何宗宪设计有限公司　项目地点：中国，香港/九龙站/圆方　摄影：Ray Lau
完成时间：2007年

The design of the first "ProNature" outlet for Asia's leading healthcare company Eu Yan Sang applies the traditional Chinese philosophy on natural phenomena, "Wu Xing". The five elements: metal, wood, water, fire and earth, have been carefully applied and interpreted in a modern sense and manner to demonstrate the brand's mission of "caring for mankind".

The design aims to regenerate Eu Yan Sang's new product line and its core focus in traditional Chinese medicine into a new and fresh image. Graphics generated from herbs related to the five elements are used on the wall and ceiling feature respectively, and two trees are planted from floor to ceiling at the shop front that serves as both a loose screen and an attractive product stand. As the name of the shop suggested, natural elements are emphasised in the whole design of ProNatural. Natural colours with warm lightings create a natural and fresh atmosphere that act as an act-on and tone up with the image of the brand and it's products. All in all, the ProNature shop presents a young and refreshing image that change people's preconceived idea towards Chinese herbs and supplements, paving new acceptance of Chinese herbs especially among the teenagers.

以传统中药为本的保健集团余仁生于香港九龙站的圆方商场开办首间"天然坊"，将中国的"五行"哲学糅合传统中医药理念，务求以清新的形象推广天然保健产品。仔细了解过"天然坊"品牌的产品理念和市场定位后，设计师决定以现代的手法将品牌造药的五行理念延续到店铺的室内设计之中。

"五行"学说是通过研究各种演化过程、功能及自然现象发展出来的一套理论，其元素分别为金、木、水、火、土；为配合"天然坊"的年轻市场定位，设计师将五行概念重新包装，研究个别中草药的特性，与传统五行元素配对，发展出一套以五行为设计基础的中草药图案，并形象化地引用到天花和货架设计中，成功赋予"天然坊"清新而富现代感的店面形象。

除此之外，设计师在店铺近橱窗的位置又"栽种"了两棵树木，从地面一直延伸到天花板，与一连串镂空在假天花的草药图案接连，树干化身成展示柜，展示重点推广的产品，再配合透明层架，达到轻盈、通透的视觉效果。为突出"天然坊"展区，设计师特别以纯白色调配衬白橡木和镜材为主要装饰材料，营造简练明亮的感觉，与周围的深木色货架形成强烈对照。传统与现代的融和，成就品牌源远流长的济世精神，令"天然坊"在众多美容和保健品牌中脱颖而出。

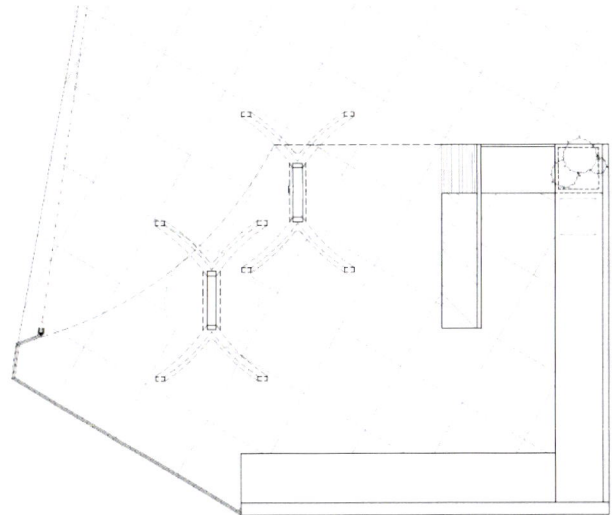

Healthy Life Natural Health Food Store

健康生活天然保健品店

Project Name: Healthy Life Natural Health Food Store   **Designer:** Design Clarity
Location: Sydney, Australia   **Photographer:** Design Clarity   **Time:** 2007

项目名称：健康生活天然保健品店 设计师：Clarity设计公司 项目地点：澳大利亚、悉尼 摄影：Clarity设计公司
完成时间：2007年

Healthy Life offers a wide variety of natural health and beauty products, vitamins, minerals and sports supplements. Having built a reputation as the leader in the health sector, Healthy Life required a new, invigorated store design to reflect the quality of their product ranges.
The design brief was to develop a memorable, new identity for Healthy Life including 2D graphics, 3D environmental branding, packaging, uniforms and a full store design concept that could be easily adapted to a wide variety of sites throughout metropolitan and regional Australia. Inspired by nature, Design Clarity developed a brand icon based around the tree motif. Crisp, green colours were specified to evoke a feeling of energy and well-being, while raw-looking finishes were adopted to reflect the honest, natural feel of the brand.
Ceilings and services are left exposed and sprayed out black as a backdrop to the basic white, green and timber palette. Adaptive re-use of familiar, natural, and earthy materials in new and less predictable ways engages the customer and allows for an environmentally aware approach to the design.
The entrance of the store features a striking timber slat shopfront that frames a glass window display, providing opportunity for dynamic visual merchandising and seasonal product displays. A bold ceiling feature draws customers into the space and departmental suspended chalkboard signage blades throughout the store facilitate easy navigation.
Market-style merchandising units including baskets, ladders and timber crates are included to allow each franchisee to create feature displays to suit their individual store requirements. Materials such as natural pine timber, zinc plated steel, chalkboard, polished concrete tile and simple pegboard are used to reinforce that familiar, comfortable and environmentally-friendly look and feel.

健康生活品牌被誉为"保健领域领军者"，旗下包括多种产品——天然保健美白品、维生素、矿物质等。因此，客户需要一个全新、活力十足的设计理念，借以用商店空间彰显其产品的多样性。根据这一要求，设计师首先构思了一个全新的标识，应用于二维图形、三维标牌、产品包装、员工制服上，打造统一的理念。空间设计受到自然的影响，绿色被大量引用进来，给人健康、动态之美，简约的装饰则突显品牌天然、保健的特点。
天花及服务区内未经任何装饰，黑色的背景为白、绿、棕等色调营造完美背景。材料使用上别出心裁——"旧料新用"的方式格外吸引顾客，同时达到了环保要求。商店入口处，木质板条结构格外醒目，玻璃橱窗镶嵌其中，用于展示季节产品。天花设计更具特色，吸引顾客走进来。店内随处可见的垂悬着的招牌为顾客在店内的"探索"提供方便。
篮子、梯子以及木条箱等用作产品展台，店主可以根据需求打造属于自己的特色。此外，天然松木、镀锌钢、黑板、抛光混凝土地砖等材质的运用更给人舒适、环保的感觉。

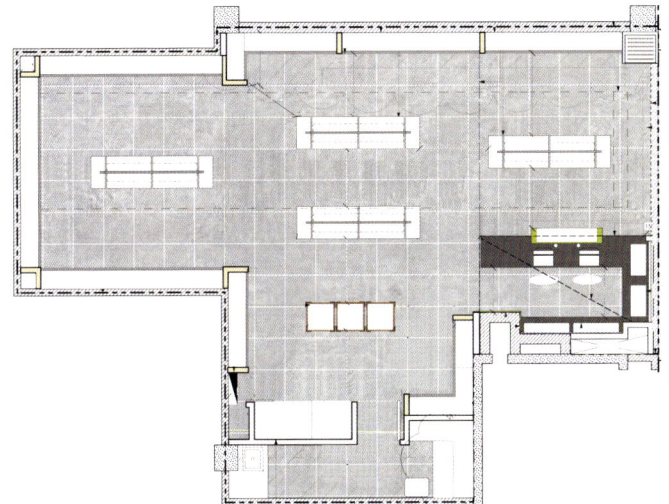

Barbie Shanghai

上海芭比娃娃旗舰店

Project Name: Barbie Shanghai **Designer:** Slade Architecture **Location:** China, Shanghai
Photographer: Iwan Baan **Time:** 2009

项目名称：上海芭比娃娃旗舰店 设计师：Slade建筑事务所 项目地点：中国，上海 摄影：伊万·巴安 完成时间：2009年

The 35,000-square-foot Barbie Flagship for Mattel holds the world's largest and most comprehensive collection of Barbie dolls and licensed Barbie products, as well as a range of services and activities for Barbie fans and their families.

Mattel wanted a store where "Barbie is hero", expressing Barbie as a global lifestyle brand by building on the brand's historical link to fashion. Slade Architecture were charged with the design of everything. Barbie Shanghai is the first fully realised expression of this broader vision. It is a sleek, fun, unapologetically feminine interpretation of Barbie: past, present, and future.

The façade is made of two layers: molded, translucent polycarbonate interior panels and flat exterior glass panels printed with a whimsical lattice frit pattern. Mattel collaborated with designers at BIG, the Branding and Design division of Ogilvy and Mather, who created the final exterior frit graphics. The two layers reinforce each other visually and interact dynamically through reflection, shadow and distortion.

For the new façade, they combined references to product packaging, decorative arts, fashion and architectural iconography to create a modern identity for the store, expressing Barbie's cutting-edge fashion sense and history.

Upon entry, visitors are enveloped by curvaceous, pearlescent surfaces of the lobby, leading to a pink escalator tube that takes them from the bustle of the street, to the double-height main floor.

The central feature within the store is a three-storey spiral staircase enclosed by eight hundred Barbie dolls. The staircase and the dolls are the core of the store; everything literally revolves around Barbie.

The staircase links the three retail floors:

The women's floor (women's fashion, couture, cosmetics and accessories). The doll floor (dolls, designer doll gallery, doll accessories, books). The Barbie Design Centre, where girls design their own Barbie is on this floor. This activity was planned by Chute Gerdeman Retail and designed by Slade. The girls floor (girls fashion, shoes and accessories).

上海首家芭比娃娃零售店占地3252平方米，拥有世界上数量最多、样式最全的芭比娃娃以及其旗下许可生产的其他产品。此外，店内还为芭比迷们以及他们的家人提供一系列的服务。

客户要求设计师为他们打造全部，通过品牌与时尚与生俱来的关联展现其国际化风格，是芭比产品成为空间真正的主角。Slade的设计包括芭比的过去、现在和未来，营造一个新潮、趣味十足的空间。

店面外观由透明的聚碳酸酯模切板和印有无规则格子图案的平滑玻璃板构成。两层结构在视觉上相互补充，通过光影的变幻，形成强烈动态感。

在外观设计上，Slade将产品包装、装饰艺术、流行样式以及建筑图片相结合，赋予商店现代化身份，借以突显芭比的顶尖时尚感，同时不失表现其历史特色。

走进店内，最先到达大厅，弯曲、灯光闪烁的表面格外引人注目。继续前行，走进粉色的电梯即刻远离街道的喧嚣，将顾客带到店内的商品区。

店内的主要特色是一个3层的螺旋式楼梯，内附800个芭比娃娃。楼梯和娃娃是商店的核心，而一切设计又全部围绕着芭比展开。

楼梯连接三个零售楼层：娃娃楼层（娃娃、玩偶设计师廊、娃娃配件及书籍）；芭比娃娃设计中心，在这个楼层，女孩可以设计自己的芭比娃娃；女孩楼层（女孩时装、化妆品及配件）。

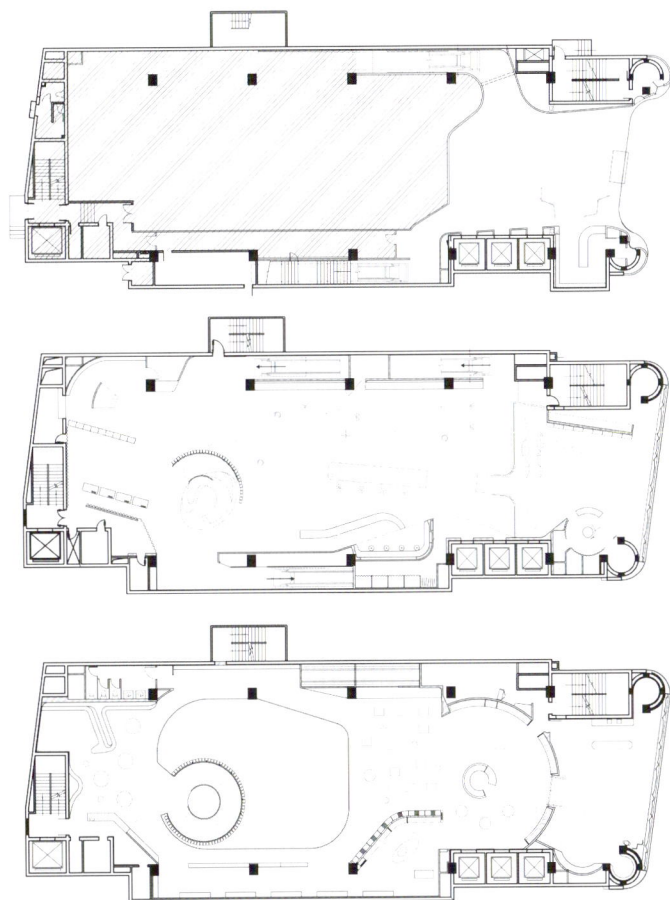

WEEKDAY MALMÖ

周末假日时装店

Project Name: WEEKDAY MALMÖ   **Designer:** Electric Dreams   **Location:** Malmo, Sweden
Photographer: Fredrik Sweger   **Time:** 2006   **Area:** 1000 sqm

项目名称：周末假日时装店 设计师：Electric Dreams 设计公司 项目地点：瑞典，马尔默 摄影：弗瑞德里克·斯维格
完成时间：2006年 面积：1000平方米

Weekday is a visionary, young clothing company. It has grown rapidly from being a small hobby project, open on weekends only, to five full-size stores, much thanks to their Cheap Monday success. The retail chain has, along with Monki, recently been sold to H&M. Weekday wants to grow without repeating itself. They want each of their stores to be unique and every project starts as a blank canvas.

A 1000-square-metre fashion store in central Malmo opened on October 26th 2006. The entrance level hosts in-house jeans brand Cheap Monday and top floor contains Swedish and international designers, second hand garments, and shoes. The top floor also houses a screen-printing studio, a fully equipped industrial laundry for dying jeans, a DJ stand, and a sewing studio.

The designers like to think of interiors as landscapes and structures. At a first glance, the interior is a maze. The angular mirrors are creating spatial confusion. It's skewed and chaotic, lots of edges and corners. But as soon as you start to explore the shop you realise that all functions are seamlessly integrated and that the green ramps form a distorted honeycomb grid that is projected on the inner ceiling. The structure is creating a non-linear flow, an unusual way of moving in a shop. There is no obvious route from dot A to dot B. It's a scripted journey from folded jeans to shoes to fitting rooms to checkout.

They think of the store as an abstract tree, with its supporting root system at the Cheap Monday level and its branches being the chrome-plated giant clothes hangers at the external designers' level, all with a pinch of disco flavour.

周末假日是一家年轻、活力十足的服饰公司，从最初只在周末营业的小店铺发展到共有五家分店的连锁时装店。这一成功在很大程度上归功于"便宜星期一"品牌的蓬勃发展，最近其产品全部被瑞典时尚公司H&M收购。客户要求，他们的每一家店铺都应该具备自己的特色。

周末时装店位于马尔默市中心，占地1000平方米，于2006年10月26日开业。一楼是瑞典知名牛仔品牌"便宜星期一"专卖店；二楼是来自各国设计师的旧货服饰店；顶层包括网印工作室、牛仔服饰染洗店、DJ台以及缝纫室。

服装店内部像是由不同的景观或不同的结构拼接而成。第一眼望去，更如同一个迷宫：角状的镜子使得空间看似一片混乱，到处是歪斜、无序的边缘与角落。然而，走进去，会看到不同的景象——所有的功能区流畅而连续；绿色的坡道恰似被揉搓过的蜂房，凌乱的格子结构反射到天花板上。在不同区域之间走动——从牛仔区到鞋区到试衣间再到收银台，如同一次事先规划好的旅行。

服装店又如同一棵树，一层和二层内镀铬的高大衣架构成了"枝叶"。

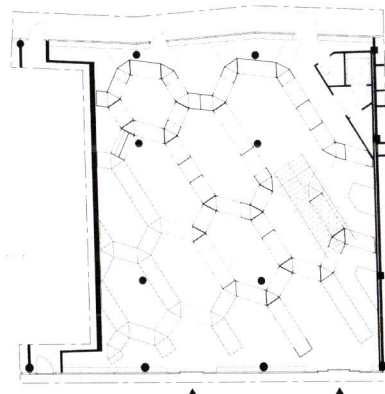

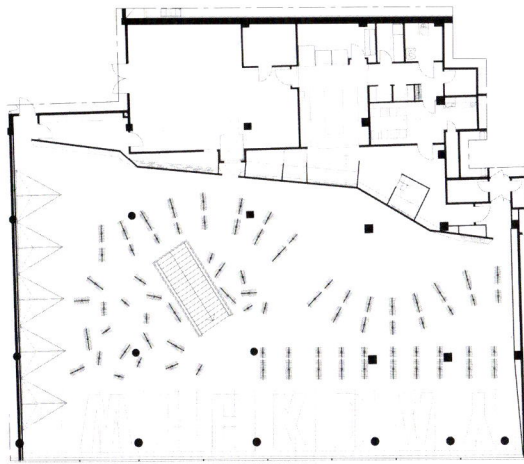

Monki Stockholm

曼奇品牌店

Project Name: Monki Stockholm **Designer:** Electric Dreams **Location:** Stockholm, Sweden
Photographer: Fredrik Sweger **Time:** 2007

项目名称：曼奇品牌店 设计师：电动梦想设计公司 项目地点：瑞典，斯德哥尔摩 摄影：弗瑞德里克·斯维格
完成时间：2007年

Monki is a new chain of fashion stores for girls, carrying its own moderately priced clothes line as well as selected designer brands. The Monki concept is based around a story about a world of creatures; hatched from chimneys, grazing sea shells. The Monkis dwell in a world made up of both the magical and the ghastly. The Peacock Fields, the City of Oil and Steel, and the Rosehip River are all landmarks in this suggestive scenery. When you walk into the store's abstract forestlike landscape, you step into a part of the Monki World.

The interior is an attempt to add to the myth of this mystic being. Details and references are to trigger imagination; entering a fitting room is stepping into a hollow oak, clothes rails are glowing tree branches, and accessory displays are shiny chrome-plated mushrooms. The Monki character is not only featured in logos, printed patterns on garments; it is also represented as stuffed animals. To create a flexible, detached world adjustable to any given store layout, a catalogue of modular key elements was designed: the Monki Alphabet. The basic grammar for this scale 1:1 Lego is a c/c 1200 modular sequential system of loops. Every loop is an unbroken line, a dissolved horizon where the floor climbs up on the walls and the ceiling reaches down to meet the floor. The elements can be combined in numerous ways within each loop, to 3D-transform generic retail space into a highly differentiated spatial experience.

In its first year, Monki opened 11 stores throughout Sweden, challenging the notion of generic low-price fashion retail with an exclusive boutique-style interior. Monki rather puts their marketing budget on a high profile store design than on full-page spreads in local newspapers. The Monki Götgatan store was the seventh in the Monki 1.0 series and opened in late March 2007. The current concept is the three-dimensional interpretation of the Monki Wetlands. Next years' establishments will feature another place in the Monki World: Monki 2, the City of Oil and Steel, an all-new interior concept.

曼奇是一家女性服饰连锁店，拥有自己的服装生产线，同时兼营一些精选设计师品牌服装。室内空间设计围绕着一群小生灵的故事展开，服装"栖息"在这个虚幻与现实并存的小世界里。

设计师通过细节增添空间神秘感，激起顾客的想象。走进试衣间，如同进入被挖空的"橡木主干"，衣架是"树枝"，配饰则构成金光闪闪的"蘑菇"。

商店的特色不仅仅体现在标识、服饰上，毛绒小玩具也格外吸引眼球。为打造空间灵活性，设计师运用了可调节循环模块结构。每一个小空间都构成一个连续的整体——地面一直"爬升"到墙壁上，屋顶"顺延"下来，与地面相接。这些模块连结在一起，便为顾客营造了一个与众不同的购物体验。

曼奇第一年内共在瑞典开设11家店面，开创了"独一无二的室内购物环境，低价精品时装零售"的理念。曼奇格外注意店内的设计，相对于在地方报纸上整版宣传自己，他们更喜欢把钱花在高品位店内氛围营造上。现今，他们更是秉承"以三维形式诠释室内空间设计"的理念，2008年将要开设的新店，将会以这一原则为宗旨。

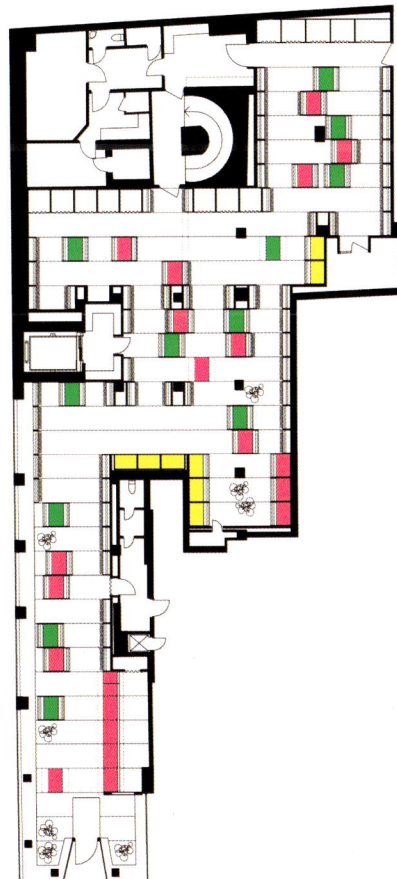

MONKI 1: Forgotten Forest

MONKI品牌零售店–遗忘的森林

Project Name: MONKI 1: Forgotten Forest　**Designer:** Electric Dreams
Location: Linkoping, Norrkoping, Stockholm, Vaxjo, Eskilstuna,Gothenburg, (Sweden)
Photographer: Fredrik Sweger　**Time:** February 2008- March 2009

项目名称：MONKI品牌零售店–遗忘的森林　设计师：电动梦想设计　项目地点：瑞典　摄影：弗德雷里克·斯维格
完成时间：2009年

Once upon a time, an unexpected chemical reaction in an old weavery released a batch of plump black Monkis into the sky. Soaring high from the still-smouldering chimney, the tiny creatures floated into the illusory universe known as the Monki World. Appearing simultaneously with the Monkis was a new fashion label and a retail concept for girls; combining graphics, goods, and store design to form an overall story. When you enter a Monki store, you step into a part of the enchanted Monki World.

The Monki World is a story about a parallel universe inhabited by little black creatures with dual personas, born in the derelict City of Oil and Steel. Monkis are cute and friendly, but also evil and deceptive. Their world is made up of places as surprising and ambiguous as themselves; part magical and part ghastly; stunning beauty alongside revolting ugliness.

This story is the foundation for Monki's all-embracing retail concept. The story is not only communicated as printed patterns on garments, graphics, websites, and store designs; the Monki World is represented even in the smallest details, such as accessories, shopping bags, price tags, and receipts.

Monki has several interior concepts running simultaneously. Every concept is inspired by a different part of the Monki World. So far, the mysterious Forgotten Forest and the powerful City of Oil and Steel have been launched. Future concepts will portray other areas of the vast Monki landscape. Ultimately, all the stores together form an entity that combines recognition, repetition, and surprise.

The all-new concept for Monki, "the City of Oil and Steel," is a surreal world full of chemical left-overs, broken skyscraper parts, and giant beaded necklaces. The floor is glossy black flowcrete, like wet neon asphalt. The wall interior is an abstract collage-city of broken skyscraper parts of different scales. The walls are completely mirror-clad, creating an illusion of horizontal and vertical free-floating skyscrapers, serving both as clothes rails, accessory displays, and shoe displays. Small accessories are sold in 7 cm perspex balls. Coloured sprouts are springing up from a poodle of poisonous chemical residue, to form product displays.

从前，在一个古老的纺织厂里产生了令人意想不到的化学反应——一群毛绒绒的小生灵被释放出来，并直飞向天空。他们在静静地烟囱上盘旋，最后落到了一个梦幻十足的宇宙——Monki世界。从此，Monki时尚女装品牌便诞生了——平面图形、商品以及室内设计共同勾画了一个完整的故事。走进店内，就仿佛走进了令人迷幻的Monki世界。

Monki世界里上演着一群来自废弃石油钢铁城的黑色小精灵们的故事。他们聪明伶俐、和蔼友好，他们又无恶不作、欺诈成性，他们的世界既令人惊讶又让人琢磨不透，时而神秘莫测，时而真切现实——魅力与丑陋并存。

这个神奇的故事就构成了Monki品牌零售店的设计基础，服饰上的印制图案、平面图形、网站、空间设计以及配饰、购物袋、价签、收银条上，到处都是Monki世界的印记。

实际上，店内设计理念并非只有一个，而是由所有理念构成一个整体。每个理念代表着Monki世界的一部分。到目前为止，"遗忘的森林"、"石油钢铁城"都已经成功问世了。

全新的"石油钢铁城"营造了一个由化学物质残留、破败的建筑以及巨大的珠子项链组合而成的大环境。地面上铺设着黑色抛光物质，如同流动的沥青；墙壁上是一幅破败的建筑部分拼贴而成的抽象画，满墙的镜子共同构建了一个虚幻的流动大厦，既可于挂衣服，又可用于展示饰品盒鞋子；小饰品摆放在7厘米大长的塑胶盒子里，供顾客选购；颜色各异的"小苗"结构从一片有毒的化学残渣中"挺身而出"，形成产品展示区。

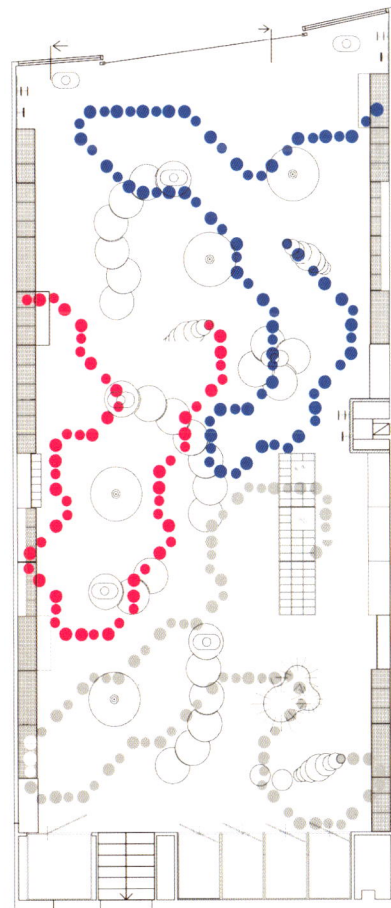

LEVI'S BB Barcelona Summer

李维斯

Project Name: LEVI'S BB Barcelona Summer   **Designer:** Jump Studio   **Location:** : Barcelona, Spain
Photographer: Go Sugimoto   **Time:** 2007

项目名称：李维斯 设计师：Jump工作室 项目地点：西班牙，巴塞罗那 摄影：杉本摄影 完成时间：2007年

This is another stand for the fashion fair Bread & Butter in Barcelona. Here the designers were asked to maintain the overall creative platform of 'The Night' but create something antithetical to the previous design. The brief was: 'Extremes and Everything In-Between'. This was to tap into particular product stories with the blackest of black washes to the whitest of white jeans.

The inspiration came from the Pantone reference cards where a colour range is demonstrated from one end of the spectrum to the other. They chose the colours of the Night, from the luminous white of the moon through to the blackest of blues for the deep, dark night sky. Rather than create the secretive, discovered spaces of the previous stand, here they created 360° open views of the stand from the centrally located bar, itself circular and lit, symbolizing the moon, with the product stories radiating out from this element on laser precise strips of colour.

Jump工作室应邀为巴塞罗那面包黄油时尚女装展设计展台，要求打造一个以"夜晚"为主题的创意场景。设计以"极限及相关事物"为理念，中间融入特殊产品讲述故事——黑色中的最黑色衣服到白色中的纯白色牛仔裤。

色彩灵感来源于潘通色卡——一个色域的颜色从一个极限到另一个极限完全被展示。他们选择夜晚的颜色，明亮的纯洁月色到深黑色。设计师并没有打造神秘感，相反，他们创造了一个完全开放的空间，从中央的酒吧可以将展台的景象一览无余：作为标识的月亮，以及周围摆放的产品全部映入眼帘。

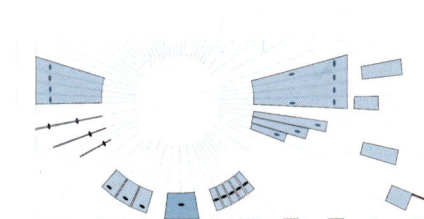

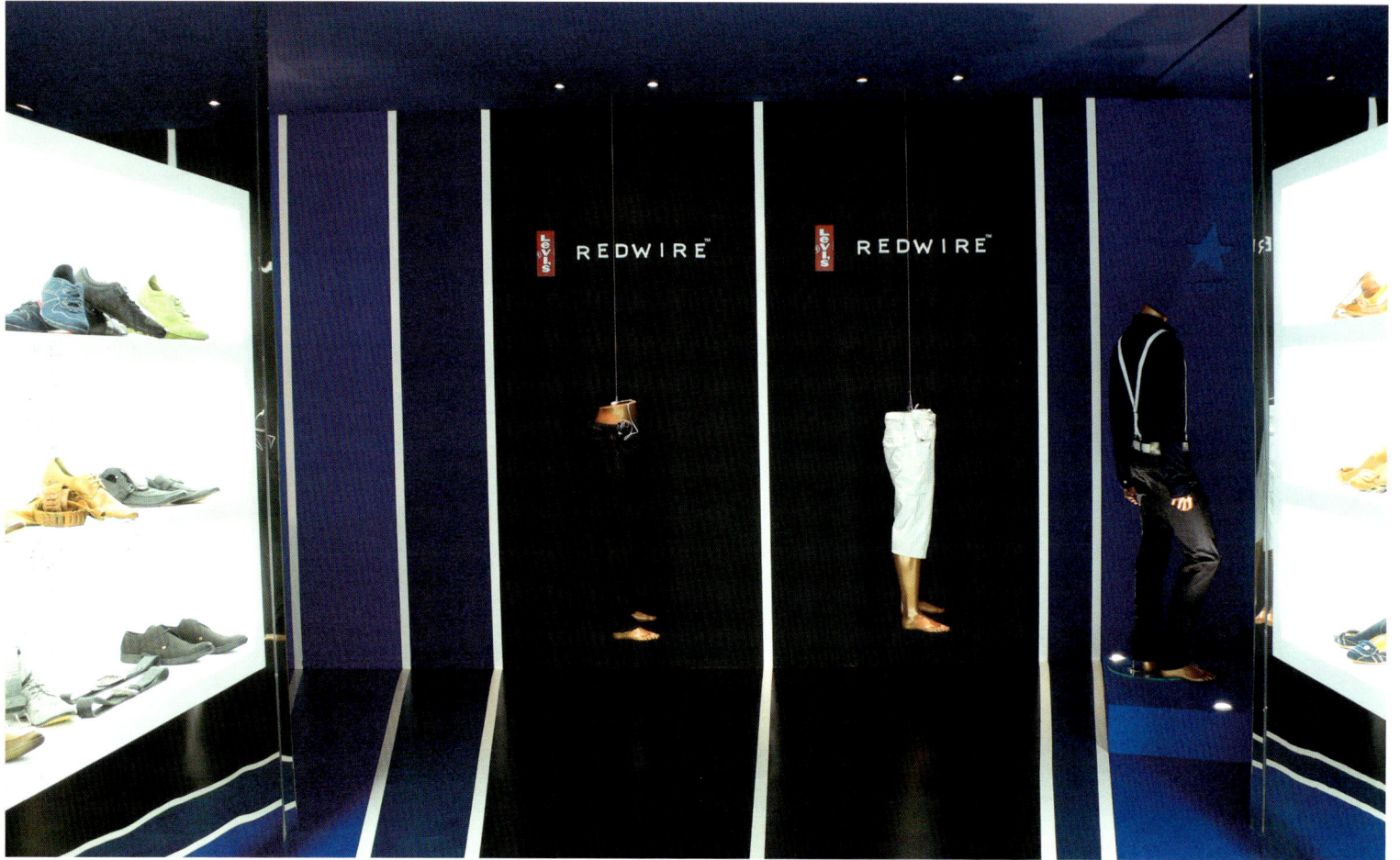

Forum Duisburg

杜伊斯堡广场购物中心

Project Name: Forum Duisburg   **Designer:** Chapman Taylor Architects
Location: Duisburg Germany   **Photographer:** Chapman Taylor Architects   **Time:** 2008

项目名称：杜伊斯堡广场购物中心 设计师：查普曼·泰勒建筑师事务所 项目地点：德国，杜伊斯堡
摄影：查普曼·泰勒建筑师事务所 完成时间：2008年

Forum Duisburg is an inner-city regeneration project located in the heart of Duisburg. The team worked with T+T Design to deliver the new, successful retail environment. The numerous pedestrian accesses linking the Forum with the surrounding streets play an important part in integrating the development to the existing urban grain. The malls are designed as individual buildings to create a continuation of the urban fabric and specifically the pedestrian street.

The 106,000-square-metre development offers 1,200 parking spaces, a direct underground railway link, a children's day care centre and a range of services for people with restricted mobility. It also includes 57,000 square metres of retail space spread across four levels and comprising 90 retailers, making it the largest inner-city retail project in North Rhine-Westphalia.

One of the principal anchor stores of Forum Duisburg is a new five-level Karstadt department store with an attractive landscaped roof garden. Spacious voids at all levels allow generous views between the four retail floors and create an open environment with squares, arcades and terraces. A harmonious, naturally-lit interior combines brick and stone walls with wood, brick and natural stone flooring.

As a part of the art programme, a specially designed 65-metre-high sculpture, 'Goldene Leiter' (Golden Ladder) connects all levels of the Forum and projects 54 metres above street level, protruding 35 metres through the glass roof to create a new, distinctive landmark on the city skyline. The ladder was made from 32 tonnes of steel that were then covered with 24-carat gold leaf.

Forum Duisburg is the first retail project in mainland Europe to be awarded a BREEAM 'Very Good' rating for its impressive sustainable features. The building materials used are prevalent throughout the region. A block heating unit allows power, heat and refrigeration to be generated on the spot, reducing average total energy consumption by at least 25 percent compared to that of similar shopping centres. The 10,000 square metres lawn roof above the department store was installed to improve the microclimate of the surrounding area.

杜伊斯堡广场购物中心位于市区中心。查普曼·泰勒建筑师事务所与T+T设计合作，负责其翻新工作，以期共同打造了一个成功的、全新零售空间。多条步行通道将广场同周边的街道连接起来，将购物中心融入到所处的城市环境中。这几幢单体建筑在某种程度上延续了城市的肌理。

这一开发项目共占地10.6万平方米，提供有1200个车位的停车场，包括一个直通的地铁网、一个儿童日托中心和一些为残障人士提供的活动空间等。购物广场作为北莱茵–威斯特法伦州最大的城中零售项目，占据5.7万平方米，共为四层，包括90家零售商店。与其毗邻的是一个新建的、五层百货商店，带有屋顶景观花园。

购物中心每层都设有一个开阔的空间，带有广场、拱廊和平台等，顾客站在这里可以纵观这里的整体景象。室内格局流畅，光线充足，墙面采用砖石材料打造，而地面则采用木材、砖头和石块铺设。

设计团队运用32吨钢材、24克拉金箔饰面特别制作了一个65米高的雕塑——金云梯，在街面54米处以上"屹立"，将购物中心的所有楼层连接起来，并"冲破"玻璃屋顶，继续延伸，在城市的天际线上形成了一个特色十足的地标标识。

鉴于其环保特色，杜伊斯堡广场购物中心曾被建筑研究所环境评估处评为"欧洲大陆第一个、极佳的零售项目"。设计中大量使用了当地常见的材质，集中供热设备可以满足区域内所有的电、热和制冷等能源需求，同此类购物商场比，减少了25％的能源消耗。更值得一提的是，建筑屋顶的1万平方米的草坪为改善周边的小气候起到了至关重要的作用。

Sexta Avenida

第六大街购物中心

Project Name: Sexta Avenida   **Designer:** Chapman Taylor Architects   **Location:** Madrid Spain
Photographer: Chapman Taylor Architects   **Time:** 2006

项目名称：第六大街购物中心 设计师：查普曼·泰勒建筑师事务所 项目地点：西班牙，马德里
摄影：查普曼·泰勒建筑师事务所 完成时间：2006年

Situated 13 km from the centre of Madrid in a prime location, this retail centre has been refurbished by Chapman Taylor Architects. Natural lighting through a new ceiling arrangement creates a brighter interior. Customer flows are improved and vertical circulation is added around a new meeting area. Floors, balustrades and other details are simple and contemporary, using good-quality materials.

购物中心位于距离首都马德里13公里的黄金地段，其翻新工作由查普曼·泰勒建筑师事务所操刀。自然光线穿过新建的屋顶透射进来，室内更加明亮。小型广场周围新增的电梯为不断增加的客流量提供了更多方便；地面、栏杆和其他细部设计简洁，全部采用高质量材料，简约而现代。

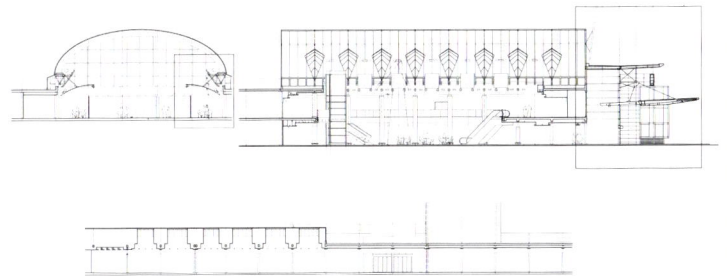

Celebrity Solstice

名至巡游艇

Project Name: Celebrity Solstice   **Designer:** 5+design project team   **Location:** Los Angeles, USA
Time: 2008

项目名称：名至巡游艇 设计师：5+设计 项目地点：美国，洛杉矶 完成时间：2008年

Celebrity Solstice, part of Celebrity's Solstice-class fleet and the first of four vessels, has an exceptional range of guest-inspired services and amenities. Inspired by world-class resorts, Solstice offers its 2,850 guests the highest level of luxury, comfort and choices within the most stylish and varied public spaces at sea. 5+design drew on the inspiration of world-class resorts to assist Royal Caribbean in transitioning from a brand concept to a physical reality, creating a luxury product on the sea that will rival most land-based resorts.

As part of the Solstice design team, 5+design was responsible for planning the retail armature – the main spine of the ship, which includes the Fortunes Casino, Tasting's Bar, Art Gallery and the retail spaces of this next-generation ship. The Fortunes Casino was designed to evoke the luxury and grandeur of Europe's finest gaming resorts in a warm contemporary architectural language. It features a huge elliptical table gaming salon, as well as an exciting array of slot machines and other automated games of chance. Fortunes' elegant lounge with its luminous etched glass bar and rosette-coffered, gold-leafed ceiling offers both players and spectators a stake in the action.

The Art Gallery was inspired by light, transparency and interaction. Works of art are suspended on movable glass panels that open onto the deck 5 Galleria Shops and Tasting's Bar, a wonderful wine and aperitif venue. The Gallery also forms a balcony overlook to the boulevard shops below on deck 4. It is flooded with light coming from portside windows opposite the balcony. The unique and flexible design allows the gallery to be reconfigured for special events and presentations. Guests can easily meander through the space to the adjacent shops or enjoy a glass of wine while browsing the artwork, or relax on the casual seating at the balcony overlooking the boulevard shops and artwork.

The open and spacious design displays the wide variety of shopping and entertainment opportunities that can be enjoyed with new discoveries at every visit. This ship offers an upscale cruising experience with fine dining, exceptional personalized service and a great attention to detail throughout.

名至作为集团名下四艘豪华游艇之一最早面世，内部设施齐全。其设计灵感来自于世界级度假村，可容纳2850名游客，并为其提供高标准奢华服务，在海上营造一个集各种空间于一体的时尚环境。

作为名至设计团队的一部分，5+负责商业空间的规划设计，包括娱乐城、酒吧、画廊及商店。其中，娱乐城在风格设计上旨在运用现代化的建筑语言，打造欧式的奢华与庄严。高大的椭圆形游戏赌桌区以及其他设施构成了空间的主要特色。典雅的休息室、灯光闪烁的玻璃吧台以及带有花饰和金叶的天花板为人们营造了一个与众不同的氛围，让人乐享其中。

画廊的设计受到灯光、透明感以及互动感等因素的启发。艺术作品悬垂在可移动的玻璃板上，朝向第五甲板的艺术品商店以及酒吧。此外，画廊自身形成了一个露台，在这里可以遥望远处的商店景象。平台的对面是高大的窗户，阳光透射进来，整个空间备感暖意融融。独特而灵活的设计赋予画廊更多的功能，可用于举办特殊活动或展览等。客人们可以从这里走到邻近的商店或是到酒吧里小饮一杯或是来到平台上晒晒太阳，欣赏店内的精美艺术品。

开放而开阔的空间不仅仅能够满足个人多样化的购物和娱乐需求，而且每来一次都会让人耳目一新。美味可口的食物、特色十足的个性化服务以及精心打造的细节必定带来一次高品位、高质量的巡航体验！

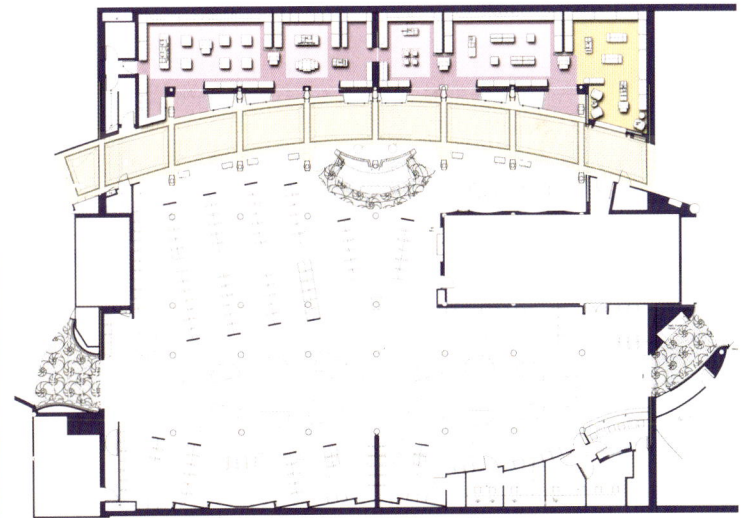

Gourmet

美食购物广场

Project Name: Gourmet   Designer: HEAD Architecture and Design Limited   Location: Hong Kong China   Time: 2005

项目名称：美食购物广场　设计师：HEAD建筑设计有限公司　项目地点：中国，香港　完成时间：2005年

This 15,000-square-foot food hall is situated in the basement of Causeway Bay's Lee Garden fashion district. The design fully utilises the dramatic impact of a five-storey sunlit atrium to illuminate the central part of the new store. This is given over to five individually designed concession 'pods', located in front of five equally spaced structural columns.

The rest of the store has been carefully planned to create a flowing but logical shopping experience in calm neutral colours, with an additional 'food street'. The 'food street' offers a variety of international lunchtime meals. This long refrigerated display counter snakes around the outer wall of the shopping centre and presents a variety of food. Taking this French fine dining theme, all departmental signage were in bilingual English and French.

By facing the preparation areas toward the customer, a better staff/customer relationship is created. The planning strategy promotes interaction, displays hygiene and shows culinary artistry.

A combination of ambient and focused lighting balances the functional and aesthetic requirements of the store. Spotlighting is focused over displays, while a variety of different light effects light the overall space.

这一占地1394平方米的地下美食购物广场坐落在香港铜锣湾利园时尚区内。设计师充分利用了来自地上五层中庭的日照光线，将广场中心区域照亮。此后，他们又围绕这个中心区，设置了五个距离相等的圆柱，并在其前方"穿插"着五个豆荚形状零售店。

广场的其余空间采用恬淡的中性色彩装饰，借以打造"流畅而条理清晰"的购物体验。一条"食品街"也布置其中，提供来自世界各地的午餐食品。长长的冷藏食品柜台在外围空间，蜿蜒环绕，售卖各种食品。此外，在这里法式餐饮作为主题，广场所有的标识牌上都标示着英、法两种语言。

"备货区"面向顾客设置，便于促使员工和顾客之间能够很好地交流。设计策略往往提升了人与人之间的交流，展示了健康卫生的烹饪艺术。

环境照明和聚光照明的共同运用平衡了广场内的功能和审美需求：环境照明主要用于照亮整体区域，而聚光照明则集中为展示的食品照明。

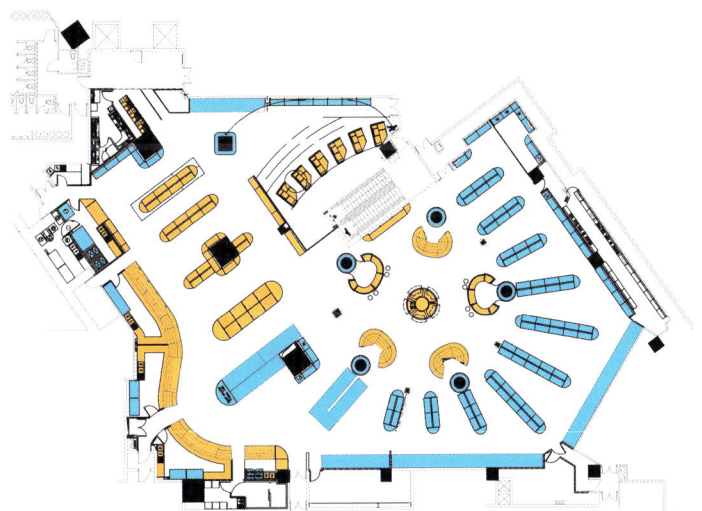

Only Glass

玻璃

Project Name: Only Glass   **Designer:** Robert Majkut Design studio   **Location:** Poznan, Poland
Photographer: Maciej Frydrysiak   **Time:** 2006

项目名称：玻璃 设计师：罗伯特·马库特设计工作室 项目地点：波兰，波兹南 摄影：马切伊·弗莱德夏克
完成时间：2006年

The place is, most obviously, made almost entirely out of glass (around 70% of the surface). Glass walls, glass ceilings, glass floors, glass staircases, glass railings and glass furniture – they all exemplify what a multi-functional material glass of all shapes and sizes is, that it can be used in a variety of surprising applications and combinations, and last but not least, that it not only an elegant, but also a pleasant material. The initial project constituted a basis for verification of the ideas and real possibilities, just before establishing the design of the Place Standards which for the designer, investor and contractor is a tool helping to create subsequent places of this kind. It denotes the only-glass trademark identity, its projects and development plans, localization and functional assumptions, materials used, shapes and finishing touches, and it serves the purpose of a consequent process of identity creation.

In order to consequently and consciously promote the only-glass trademark a sizeable Corporate Identity has been invented. Thanks to this, all the forms of the sign visible for the receiver are used in accordance with earlier specified standards as far as colours and typography are concerned. Promotional material and office stationery, or the signage inside the studio and on the elevation of the factory, have been designed according to specified patterns and requirements with moderate aesthetics dominated by yellow and black colours supplemented with grayish and violet hues.

设计师将玻璃作为最主要的元素，打造了一个几乎完全透明的空间——玻璃墙壁、玻璃屋顶、玻璃地面、玻璃头提、玻璃栏杆、玻璃家具——展现了玻璃作为一种多功能的材料的创意组合，典雅而舒适。这一设计不仅将原始的想法和可能性变成了现实，同时也为这一类空间的打造制定了标准。

为进一步推广公司的品牌，设计师根据之前的颜色和字体打造了一个巨大的标识结构。宣传材料、办公室用品、工作室的内部以及工厂的电梯完全按照特定的图案及要求设计——黑、黄两色为主，同时配以灰色和紫色。

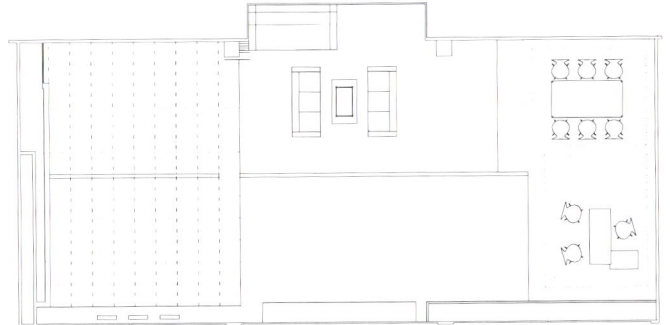

Rocawear Mobile RocPopShop

Rocawear移动服饰店

Project Name: Rocawear Mobile RocPopShop **Designer:** d-ash design **Location:** Mobile unit, various locations **Photographer:** Frank Oudeman **Time:** 2008

项目名称：Rocawear移动服饰店 设计师：德–阿什设计 项目地点：世界各地 摄影：弗兰克·欧德曼 完成时间：2008年

When rapper Jay-Z wanted to launch his new premium apparel line for his Rocawear brand, d-ash design developed a mobile lounge and retail shop within a 53-foot tractor-trailer (1000 square feet). The concept was to create a Jay-Z inspired lounge and shop that would bring his new apparel line directly to his customers.

Supporting the mobile theme, display racks were designed as travel trunks, covered in patent leather and lined in suede. Plush materials and fabrics were utilized throughout the mobile unit to offer a luxurious shopping experience.

The mobile shop will travel around the country in 2009 and pop up at concerts, sporting events and in city centers. Customers will be able to hang out, shop and listen to music.

This is a new paradigm for shopping. It is both a store and a lounge. It has no permanent location. Tuesday it can be in Philadelphia and Wednesday it can be in Boston. The shop is a mobile marketing/brand experience.

Jay-Z，顶尖说唱歌手，意欲为他的全新品牌服饰生产线—Rocawear打造一个零售空间，使其产品直接面向顾客。德—阿什设计秉承这一理念，创造了一个面积为92.9平方米的移动休息室和商店。

为强调"移动"的理念，展示架被设计成旅行箱的形状，外观采用黑漆皮装饰，里衬则采用小山羊皮打造。毛绒材料和织物被大量运用，在整个空间随处可见，营造了奢华的购物氛围。

移动商店将于2009年在整个国家巡回，在音乐会、运动会以及城市中心出现。顾客们可以在购物的同时，听听音乐或出去逛逛。

这一设计开创了购物新模式——既是商店又是休息室。没有固定的地点，周二在费城，周三在波士顿。

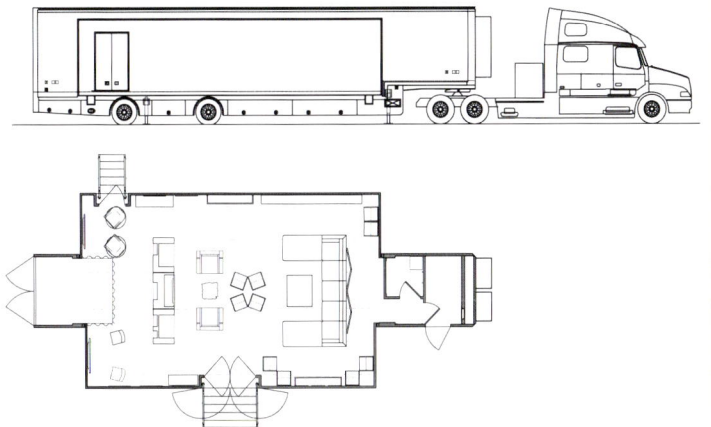

SENSORA

森索拉

Project Name: SENSORA   **Designer:** Design Clarity   **Location:** Sydney, Australia
Photographer: Design Clarity   **Time:** 2009

项目名称：森索拉 设计师：Clarity设计 项目地点：澳大利亚，悉尼 摄影：Clarity设计 完成时间：2009年

At Sensora there is a focus on service, indulgence and ultimate luxury. With a series of individual beauty treatment rooms, discreet 'microzones' for fast-turnaround mini facials, a 'skinbar' for product dabbling and makeovers, a comfortable communal hairstyling area and bespoke retail product display zones, Sensora offers the complete package. The store environment is sleek and minimal, restrained and sensual — answering the brief for the creation of an evocative space for sensory retailing.

Inspired by the science of skincare, Design Clarity created a distinct brand mark for Sensora reflecting the molecular makeup of skin cells. This logo device has been manipulated into a fragmented pattern and used as a recurring motif throughout the interior in blonde timber screening devices, sculptural wall pieces, mirrored panels, three-dimensional ceiling planes and subtle light features. The built environment and graphics have a consistent thread, which continues through the custom-made packaging concepts, uniform design, ticketing and even product shelf talkers. The shopfront design extends this scattered fractal pattern to the store façade across the frameless glass, inviting and intriguing customers to enter under a portal of pure white vitrified ceramic mosaic tiles.

Finishes have been selected to reflect a natural minimal aesthetic, American Oak flooring, white leather, mirror, low VOC paints, and white solid surface joinery. White on white walls and gloss on matt reflective signage elements add another layer of subtle texture to the interior and reinforce the Sensora brand. The day spa layout covers two adjacent retail tenancies and is set out at a 45-degree angle creating a more dynamic sense of space and allowing the display units and half height merchandise walls to face the customer on approach.

Design Clarity partnered with the Sensora directors and together delivered a unique, pure example of sensory experiential retailing.

森索拉注重服务质量，营造极致奢华。集单独的美容室、产品试用区、化妆区、美发区、产品展示区于一身，森索拉营造了一个简约而不失时尚风格的氛围，满足了顾客购物过程中的感官体验。

设计师从皮肤护理学科中获得灵感，为其打造了一个特色十足的标识，用以从分子结构角度展现皮肤细胞的构成。更值得一提的是，这一标识又被转换成零散的小图案用来装饰室内空间，木质屏风、墙壁、带镜面的板材、三维天花板以及照明设施上，随处可见。整体空间环境与平面装饰图形之间似乎被一条无形的线连接，包装袋、制服甚至是柜台前的服务人员全部完美的融合在一起。此外，店面的设计再次运用了不规则的图形结构，无框的玻璃结构让好奇的顾客们情不自禁地走进来。

装饰材料选择上更是以营造自然、简约美感为主旋律，美洲橡木地板、白色皮革、镜面、低挥发复合涂料以及白色饰面的实木家具别具特色。白色墙壁上配以白色装饰和高光标识元素，在增添空间质感的同时，进一步强化了森索拉的标识。日间水疗馆占据了两个零售铺面的位置，格局成45度角，为空间注入了更多的动态感，让顾客走进店内便可将商品一目了然。

设计师同森索拉有关人员共同合作，开拓了感官购物的新体验。

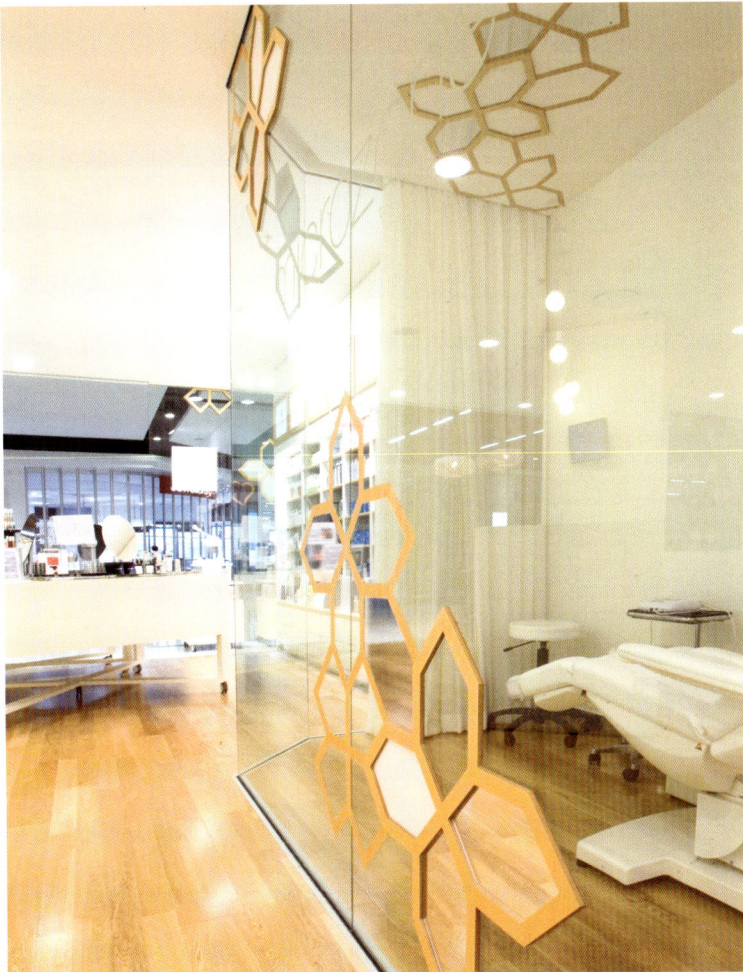

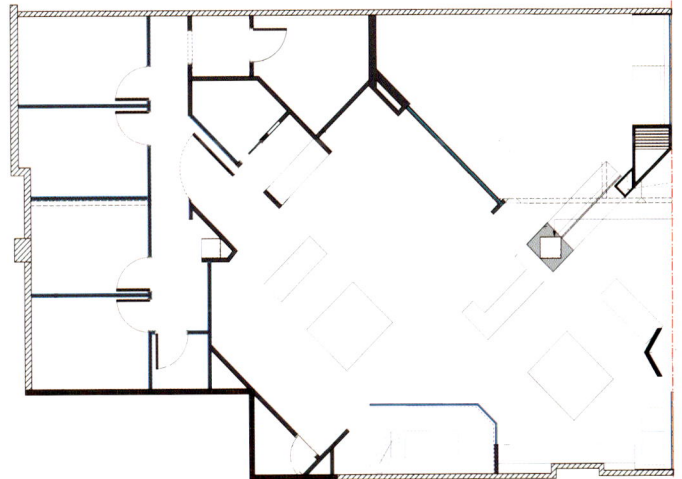

UHA Mikakuto

味觉糖食品店

Project Name: UHA Mikakuto   **Designer:** GLAMOROUS co.,Ltd   **Location:** Shanghai, China
Photographer: Nacasa & Partners, Inc.   **Time:** 2008

项目名称：味觉糖食品店 设计师：美好设计有限公司 项目地点：中国，上海 摄影：姚京摄影 完成时间：2008年

The client, UHA Mikakuto Co., Ltd., a famous long-established sweets manufacture in Japan, established its first flag shop in Shanghai, located at a popular streets' intersection.

The shop doubling as showroom has a dominating presence with the continuous huge shades above the showcases and the custom-made lighting stands covered with two-colour crystals which have been designed and are based on candies and chocolates.

While these striking lighting stands catch passengers' eyes, the classical wooden moldings dyed in chocolate colour successfully soften the whole space, being well-balanced.

The custom-made lighting fixtures above the long marble table by the window, generating different rhythm to the space, enhance their presence at night looking like floating in the air. Simple coloration of the interior flatters colourful sweets packages.

味觉糖食品有限公司是日本一家知名的糖果生产公司，于2008年在上海的黄金地段开设了旗舰店。商店同时具备展示功能，橱柜上方的巨大水晶吊灯和特制的灯架全部采用糖果和巧克力的颜色装饰，连续而流畅。高大的灯架时时吸引着行人的眼球，古典的木质结构染上巧克力色彩，使得空间更加柔和。

特制的灯具悬垂在窗边的大理石长桌上方，增添了空间节奏感，在夜晚如同悬浮在空中，神秘而梦幻。此外，简单的配色方案使得五颜六色的糖果包装盒更加突出。

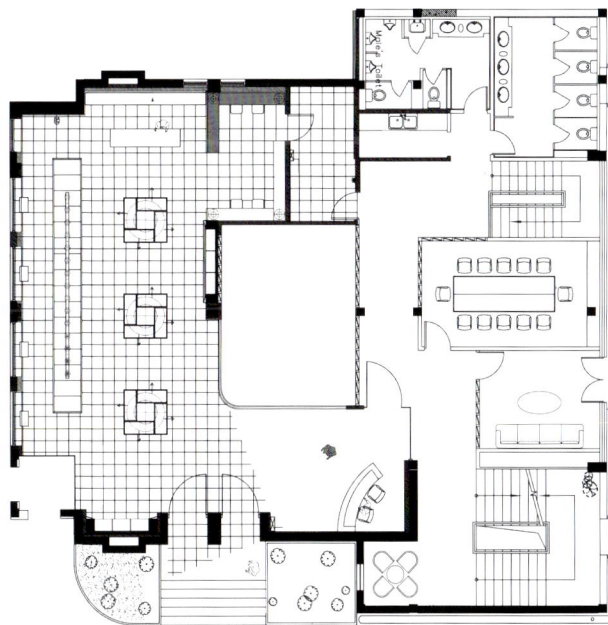

O2

O2通讯

Project Name: O2　**Designer:** JHP　**Location:** London, UK　**Photographer:** JHP　**Time:** 2008

项目名称：O2通讯　设计师：JHP　项目地点：英国，伦敦　摄影：JHP　完成时间：2008年

A four-way pitch held in April this year led to the agency being selected to partner with the company to reposition its store network, supporting its "Connected World" strategy. Decidedly moving away from being a "phone shop" to being a platform for total connectivity, the new concept is set to truly evolve the format of retailing communication devices. The intangibility of O2's content and services will be core to the store of the future.

Central to the new store design is the creation of a newly appointed O2 Guru in each store. A futuristic bar starts in the window and sweeps around to form a service area in which the O2 Guru can operate to the side of the store. The window display aims to bring the tangible benefits of being on O2 to life rather than the technology itself. Customers will be able to seek technical help from the O2 Guru on products or services or simply to discover what's new or planned to be launched.

The strategy to create a complete "connected world experience" is brought alive by the physical layout of the product ranges. Phones and technology are displayed on bespoke pieces of furniture rather than the standard approach of using wall bays. Devices are laid out on fixtures to tell complete stories, with phones connected to the best accessories on the market, including headphones, printers and speakers. All products are live to encourage customer interaction and to discover the world of O2.

The store environment seamlessly links the worlds of the contemporary home and future technology without alienating the everyday customer. Contemporary concrete floor tiles and dark timber flooring are used to define different zones within the store. Overhead, the unique ceiling features sinuous interlocking blades, reflecting the form of sound waves.

At the back of the store, is the "lounge" area. Relaxed and inviting furniture positioned around a coffee table faces a huge 65" interactive touch-screen which will show product demonstrations, gaming, video clips from O2 and from sponsored sports events. Timber flooring, textured wall-covering and lampshades bring warmth to the environment and adds to the homely feel. The lounge area will also provide a space for O2's "It's Your Community" presentations where grants are presented to local community groups.

JHP受邀为O2打造全新的电信零售理念，实现其"连通世界"的宏伟战略。设计中避免手机商店的模式，旨在打造一个以"连通"为主要理念的平台，开创零售通讯业的新篇章。O2无论是在销售的产品还是在提供的服务上，都给人一种变幻莫测的印象，当然这也会成为其未来店面的发展趋势。

未来风格的货架从窗户下面开始延伸，形成一个自助服务区。顾客可以在这里寻求技术帮助，了解店内的产品及服务或者是确知最新的或即将上市的新产品。窗口展示区的设置旨在将O2的切实利益彰显出来，同时也是对其技术的完美补充。此外，根据产品种类而设置的空间格局更是将"连通世界"的理念淋漓尽致地体现。手机及各种技术产品摆放在定制的家具上，而其他配件如耳机、打印机以及扬声器等与手机连通则安放在固定的装置上，进一步完善整体理念。这里的所有产品都会激起顾客交流互动的意愿，让他们进一步体验O2打造的世界。

店内环境是现代化家居氛围和未来技术恰到好处的融合，让每一位顾客备感亲切。现代风格的锦砖地面和深色木质地面用于区分不同的区域；头顶上风格独特的天花板上盘旋着蜿蜒互锁的叶片状结构，彰示着声波的形态。

店内的后侧是休息区。家具围绕着咖啡桌随意地摆放，面向一个65英寸的超大互动式触摸屏幕（用于展示店内的产品以及赞助商的视频）。木质地板、质感十足的墙饰以及灯罩共同营造了温馨的家居氛围。此外，在这里还预留了一个专门的空间，用以O2的特殊产品。

自助服务台的设置为顾客检索账户、购买O2活动票提供了方便，完善了他们的O2体验。此外，店内格局的设计全部以顾客的需求为出发点，为他们打造完美的购物体验。

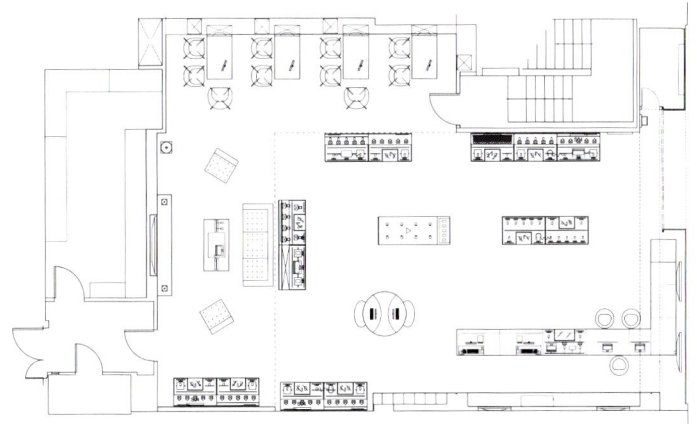

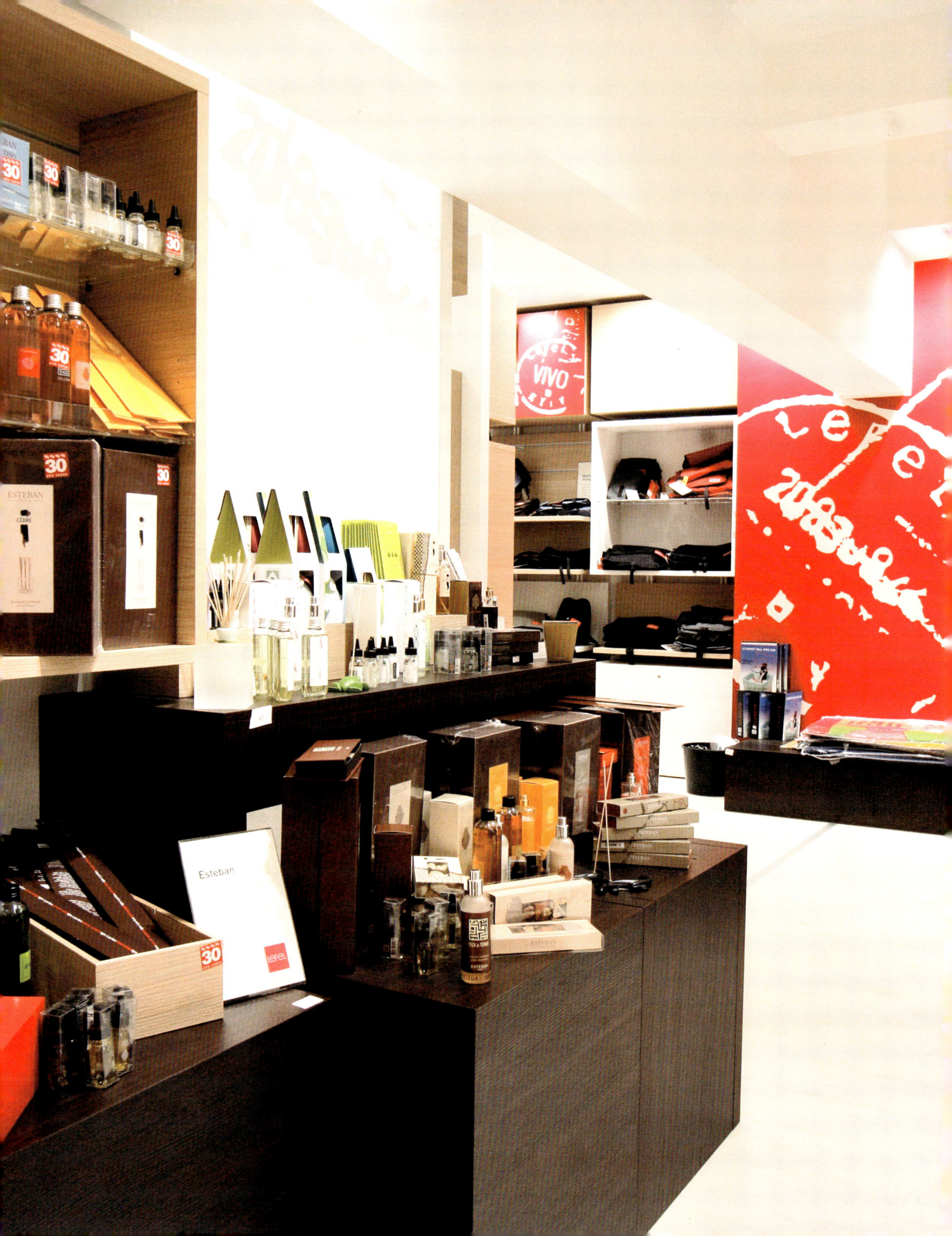
ESTEBAN
CÈDRE
30
VIVO
Esteban
TECK & TONKA

Lefel – The Whispering World

拉菲服装店-幽静的世界

Project Name: Lefel – The Whispering World   **Designer:** Crea International   **Location:** Italy
Time: 2008

项目名称：拉菲服装店-幽静的世界 设计师：Crea国际 项目地点：意大利 完成时间：2008年

The new Lefel store intends to be a place for souls where a free and adventurous spirit, sometimes eccentric, puts the most significant objects of its wandering, travel by travel. It is a place where words, sounds, scents, past and future meet and where the objects tell stories and talk to people who have chosen them. It is a mutable and essential place where mysterious boxes, as transparent and light as thoughts at times, or as sturdy as wood other times, pick together objects that whisper little, big stories to the people who are able to listen to them.

Camilla Croce, design director explains the Lefel concept: "Stongly believing that style is secondary to sentiments in order to make our surroundings more human and ethic, I thought myself how I could create an affective relationship between customers and spaces they interact with. Just because it's that precise link which pushes us to come back shopping to the same place it was given birth in. Everybody, me included, has an old wooden box at home. Full of "black and white" keepsakes such as photos, letters, postcards and experiences where we can dip into for hours."

Lefel is a dream container where anyone can listen to enchanting stories told by a free and adventurous spirit — a spirit who collected and shares them with kindhearted people who believe in affective connections with everyday life objects.

设计的主旨是打造一个让精神灵魂自由释放的"拉菲世界"——语言、声音、香气、过去与未来交融的空间，各色商品在这里讲述他们自己的故事，并将故事传递给选择他们的顾客；神秘感十足的盒子时而如同思想一样透明闪烁，时而如同木头一样坚硬顽强，他们将不同的物品"集聚"在一起，相互耳语，向那些与他们"心有灵犀"的顾客讲述他们伟大的故事。

设计总监Camilla Croce解释说："我一直坚信风格相对于情感永远处于第二位，只有这样才能够使我们周围的环境更加人性化。我的想法是如何在顾客和空间之间制造一种密切的联系，因为正是这种联系促使我们成为'回头客'。每一个人，当然我也不例外，都会在家里拥有一个古老的木盒子，装满了我们的黑白记忆——照片、信纸、明信片，让我们时不时地拿出来回味。"

拉菲就是一个梦想盒子，每个人都可以来这里聆听动人的故事——一个自由的灵魂将他们收集在一起，讲给那些热心肠的人听。

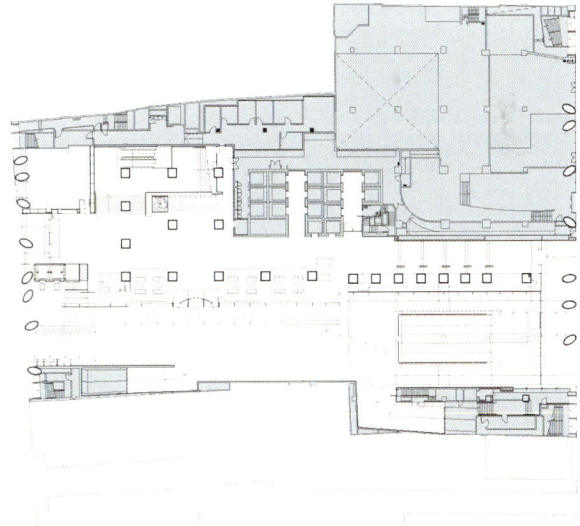

Nike Airmax 360 Exhibition

耐克

Project Name: Nike Airmax 360 Exhibition   **Designer:** Mark Lintott , Chia Yen ,Alison Yang
Location: Taipei, China   **Photographer:** Marc Gerritsen   **Time:** 2007

项目名称：耐克 设计师：林马克 阿利森·杨 项目地点：中国，台北 摄影：汤马克 完成时间：2007年

Nike invited MLD to design a promotional space for the launch their new "Airmax 360" sports sneakers. The event took place on the ground floor at the newly opened flagship Eslite Bookstore in Shin-Yi District of downtown Taipei.

The theme of this new product presentation is related to "Walking on the Air", where the light-weight and transparent structure of the shoes was the main promotional focus. With this in mind, MLD designed a space defined by screens made of semi-transparent diaphanous cloth instead of hard surfaces or walls in a conventional sense. Behind the two sides of these fabric walls, a series of predecessors of "Airmax 360" sneakers representing the shoes' history were displayed in translucent acrylic display cases seemingly floating on air. Each case was lit by bundled strands of optical-fiber cables dropped from the ceiling. Visible through the fabrics, the essence of such a display resembles the light-weight and transparent feature of the product's structure.

In the centre the main focus of the installation was a group of 6 transparent acrylic globes hung only by optical-fiber cables which also provided the light source for each shoe. The latest "Airmax 360" sneakers were exhibited inside these acrylic spheres seemingly floating in space.

MLD 受邀设计耐克新品 Airmax 360 宣传发布会的展览空间，场地设于诚品信义旗舰店的一层。

此次宣传发布会着重于产品本身的特性 "漫步于空中"，鞋底轻巧透明的结构是其主要的宣传焦点，为此设计师将其特性延伸至整体展览空间，用半透光的布幔与车缝手法来代替惯有的硬隔间以区隔整体空间。布幔两侧的后方是由钢索悬吊的长形透明亚克力展示架，陈列了此款鞋历年来的产品，由垂吊的光纤缆绳来点亮着。透过布幔可隐约看见展示的产品，其精神类似产品本身的内部结构。

而整个空间的中央悬挂着6颗透明亚克力球，球体中陈列的是最新的Airmax 360产品，借由这样的手法使得符合产品特质的穿透感与轻盈度更加具体的呈现。

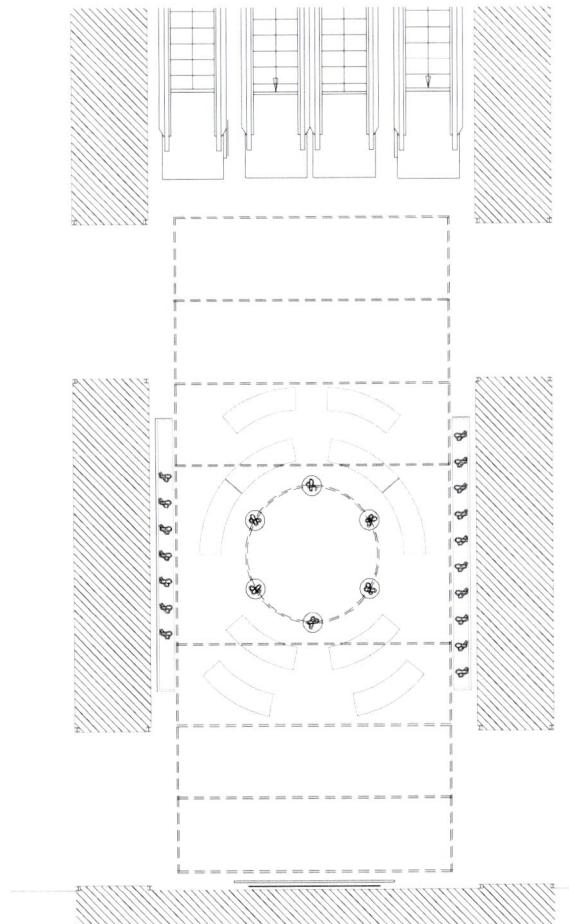

Kymyka shoes and bags, Maastricht

Kymyka鞋包店

Project Name: Kymyka shoes and bags, Maastricht Kymyka   **Designer:** Maurice Mentjens Design, Holtum **Location:** Maastricht, the Netherlands **Photographer:** Arjen Schmitz **Time:** 2009

项目名称：Kymyka鞋包店 设计师：莫里斯·门提真斯设计工作室 项目地点：荷兰，马斯特礼赫特 摄影：阿杰恩·施米茨
完成时间：2009年

Two patches of stainless steel rods, perched on top of which are artistic shoes, like colourful butterflies; an all-in-one wall unit of cupboards surrounding the space, intriguing and stylish — these are the signature style of Maurice Mentjens, responsible for the interior design of this exclusive shoe shop in Maastricht.

The shop space covers the ground floor of two city residences, the wall between which has been broken through, with a mirror-image floor plan. The houses were built at the beginning of the last century. Authentic original elements still in place include the plasterwork ceilings and two chimneys. One building still has the original windows; the other building was fitted with a large, low-silled display shop front in the 1970s.

Mentjens used the vertical lines of the premises as the basis for the structure of the design. The column that has replaced the partition wall stands exactly on the intersection of the floor plan. The steel supporting beam has been covered from floor to ceiling with mirrors. From this point, you can see the entire shop space through the mirrors. The two chimneys can be used as mirrors for customers trying on shoes. Since only the bottom 90 cm are covered in mirrors, these heavy structural features seem to float in space.

The space is entirely enclosed by low, multi-functional console tables. All the shop functions are contained in these: storage space and display cases, as well as cut-out seats and technical facilities such as the meter cupboard and heating. Only the sales counter itself protrudes into the space like a peninsula.

Another regular set of beams has been placed above the wooden console tables, a kind of minimalist wall frame forming the transition to the neo-baroque plasterwork that decorates both ceilings. These beams also embrace the room, and — through their linear simplicity — keep the irregular shape of the floor plan together, as it were. It is a surprisingly powerful design element that also has a functional side. The beams run right along the windows as well and make ideal display shelves for smaller items such as shoes and bags. The walls, ceiling and beams have been painted in the same colour and thus create a tranquil atmosphere.

鞋品"栖息"在不锈钢杆结构上，如同五颜六色的蝴蝶；店内四周的壁柜，风格时尚而引人注目——这便是莫里斯·门提真斯的独家设计风格。

商店由两幢民宅的一层打通而成。房子始建于上世纪初，一些古老的元素依旧存在，包括灰泥屋顶和大烟囱。其中一幢保留着原有的窗户，另一幢则安装了20世纪70年代的低窗台商店橱窗。

设计师充分运用垂直元素，并将其视作整个空间结构的基础。梁柱代替隔断墙矗立在空间交界处；钢梁支撑结构耸立在屋顶和地面之间，并在表面镶上镜子，这样一来，顾客就可以通过镜子看到整个商店空间；两个大烟囱也被充分利用——顾客试穿的时候把它们用作镜子。值得注意的是，所有支撑结构只有在下面90厘米处安装了镜子，因此看起来如同悬浮在半空中，创意十足。

店内满是低矮的桌案，集储藏、展示、座区、技术设备（仪表、供暖设备）容纳于一身。收银台形成了一个独立的半岛，在空间内突显出来。

此外，桌案上方安装了一组梁柱，形成了简约风格的墙壁框架，同时构成了天花板上新巴洛克样式灰泥装饰的过渡。此外，这些梁柱将整个空间围合起来，沿着窗户延伸，用于展示小的鞋包物品。墙壁、天花板以及梁柱刷以同样的颜色，营造了安静的氛围。

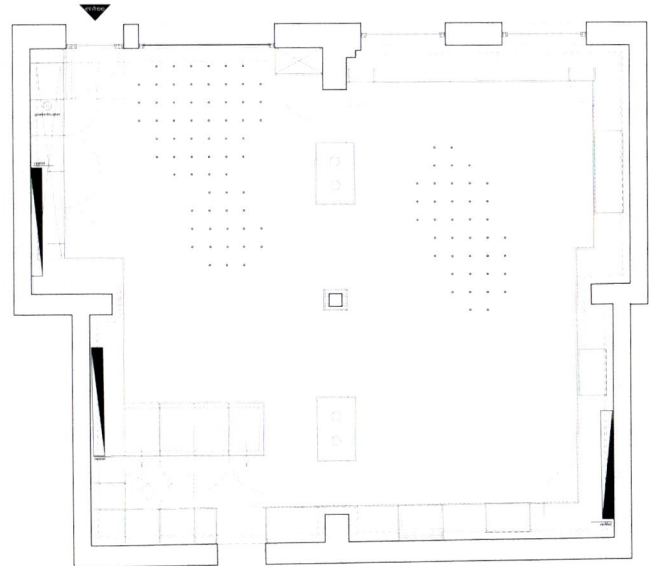

Offspring Camden

Offspring品牌店

Project Name: Offspring Camden **Designer:** Shaun Fernandes, Sarah Williams
Location: London UK **Photographer:** Mark York Photography **Time:** 2008

项目名称：Offspring品牌店 设计师：肖恩·弗门德斯 萨拉·威廉斯 项目地点：英国，伦敦 摄影：马克·约克摄影工作室
完成时间：2008年

Too many retailers slavishly try to reflect the culture of young people, witness the plethora of over-urbanised spaces, exposed brick, distressed leather sofas and graffiti.

Given the culture Offspring has always been a drive to innovate and a passion for the product; the inspiration for the design came from a need to reflect this culture, rather than slavishly following trends.

The approach has been to create a highly charged technologic landscape where nature is encapsulated in a technological playground of grass imbedded benches, cast resin trees snaking through the space and where the product is displayed on a swathe of polycarbonate 'blades' lining the perimeters of the shop. To change the pace where the product moves from footwear to apparel, they have created a system based on the language of an Air fix kit, with product replacing the model itself.

现如今，越来越多的零售商试图彰显年轻人的文化，因此导致了都市空间、砖石建筑、皮革沙发和涂鸦的过剩。而Offspring却拥有自己独特的方式，不断创新开发更多的产品；不会一味地追赶流行趋势，而是根据顾客的需求展现自己的文化。

设计师营造了一个感情丰富的背景空间。树脂材料打造的树木蜿蜒在整个空间内，增添了自然气息；聚碳酸酯材料制作而成的叶片结构环绕在店内四周，上面用于展示各种商品；常见的模特被特制的摆设结构而取代，打破了传统的展示方式。

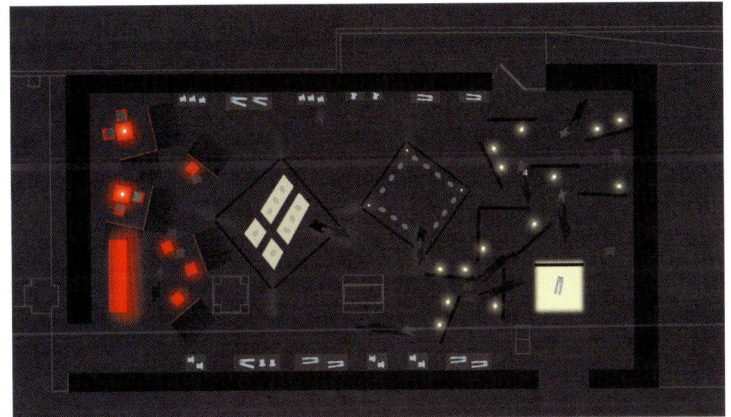

Sabateria Sant Josep

桑特·何塞普鞋店

Project Name: Sabateria Sant Josep   **Designer:** Joan Casals Pañella / CSLS arquitectes
Location: Barcelona, Spain   **Photographer:** Eugeni Pons   **Time:** 2010

项目名称：桑特·何塞普鞋店 设计师：琼·卡萨尔斯·帕奈拉/ CSLS建筑 项目地点：西班牙，巴塞罗那 摄影：欧金妮·庞斯
完成时间：2010年

The client that came from a long generation of shop assistants, had some shops but he didn't feel identified with none of the shops. He needed a different place; he wanted to personalise his sale's style, his life's philosophy and survive like that to the competence. He acquired a new business with a main goal: substituting the old business and moving to the new one when it would be finished.

The new business has two different parts: the first one which is narrower and deeper gives access from the exterior of the shop; the second one is wider inside and the exitsting programs could be laid out (warehouse on the ground floor and in the upper floor, out from the intervention zone, the services).

The design is about unifying the two spaces to suggest a continuous trajectory through the programs of the shop. Talking about analogy, they understand space as an empty bottle. If the bottle's limit was what the content contains and the content in this case were the shoes, the only solution would be solving the container's limit.

The designers suggested a furniture-partition programme that would be able to transform into function the product that the client sells and the position where it's situated; it should be able to show, contain and transform the visual perception homogenizing the space. The shapes are intertwined to improve the product's vision and they just exist some inflection points to save constructive exisiting elements. Above all, the limit becomes an interior topography that has different meanings and it's able to contain the lightning, the air conditiong, the sound and of course the shoes. It is at the same time container and content.

鞋垫的主人曾做过店员，现在拥有多个店铺，但一直觉得其千篇一律。因此，他需要一个特色十足 的空间，突出自己的销售风格及生活哲学。同时，他希望在新店建成之后便将老店取代。

新店由两个不同部分构成：一部分呈现狭长结构，空间纵深被拉长，顾客从外面直接进入；另一部分则比较宽阔，设置着仓库（一层）及服务区。设计的目的即将两部分空间统一起来，打造一个整体。他们将空间比喻成空瓶子，鞋品则是里面盛放的"内容"。如果瓶子的大小限制了其所装物体的多少，那么首要问题就是打破瓶子的限制。

最终，设计师构思了"家具作为隔断"的理念，每一双鞋子以及其摆放的位置都被功能化，在视觉上将空间均匀分割开来。不同形状的插入则使得商品更加吸引眼球，同时节省了建筑材料。尤为重要的是，空间的限制则成为了室内布局的基础，别具特色。

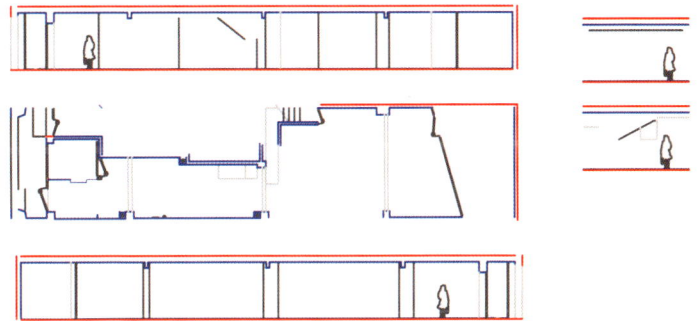

The Eyecare Company

眼睛保健眼镜店

Project Name: The Eyecare Company **Designer:** Design Clarity **Location:** Sydney, Australia
Photographer: Design Clarity **Time:** 2006

项目名称：眼睛保健眼镜店 设计师：Clarity设计公司 项目地点：澳大利亚，悉尼 摄影：Clarity设计公司 完成时间：2006年

The Eyecare Company makes a bold statement on George Street, Sydney's flagship thoroughfare, and the concept demonstrates just how effective colour and pattern can be. The heritage listed site limited the treatment to the exterior of the building and challenged the designers to grab the attention of passers-by from within. The building's structure and location was capitalised with the injection of dynamic colour and graphics, providing theeyecarecompany massive exposure. The large-scale existing interior entry bulkhead also provided the perfect canvas for the store's strong new branding concept.

The store's prominent location in the heart of the CBD called for a design solution that was distinctive but with a professional edge. It was clear from the beginning that colour and a dynamic graphic identity would play an integral role in helping catch the attention of workers as they moved through the hustle and bustle of a busy city.

The design team began the project by looking optical illusions and was drawn to Op art of the 1960s. The designers agreed that the playful nature of this movement and its art was appealing, and the preoccupation with using pattern to create illusions proved instrumental when designing the logo for the project. As a result, the circle became a repeating motif that was used in different ways to direct the eye by framing the view.

The store's entrance is a dramatic interplay of colour and form and the large feature displays combined with the mobile trolleys reflect a sense of movement. Moreover, the circular merchandising units along the Jamison Street façade are presented at different heights directing the eye along a waveform. In addition, much of the joinery uses circular cut out viewing windows to display the glasses, creating a strong focal point.

眼睛保健眼镜店位于悉尼的旗舰店大街上，设计大胆，充分展现了颜色与图案所能达到的设计效果。这个地点是遗迹保护区，对建筑外立面的处理有所限制，这就向设计师提出了挑战，他们要让设计能够利用室内吸引路人的眼球。这座建筑及其所在的地点都具备鲜艳活泼的色彩和平面图案，使这家眼镜店能过充分暴露在公众面前。建筑原有的巨大的入口隔离壁也为设计提供了完美的发挥空间。

眼睛保健眼镜店所在的位置使得设计师选择的设计方法既与众不同又具备很强的专业性。从一开始就定下了设计重点——色彩和动感十足的平面标识图案，它们将吸引在这喧嚣的城市中过往的人们。

设计团队为这个项目打造的设计以视觉幻影为起点，从20世纪60年代的欧普艺术中汲取灵感。设计师认为这些动感元素趣味性十足，这样的艺术很吸引人，而在产品标识的设计中运用平面元素营造幻影效果也证明是明智之举。于是，圆形成了空间中反复出现的一个主题，以各种不同的方式加以运用，引导着人们的眼球。

眼镜店的大门，色彩与造型巧妙融合。巨大的展示柜搭配移动手推车，也体现出动感。另外，靠近沿街外立面的圆形的售货台高度不一，呈现有高有低的波浪形。墙壁上留出窗洞，用以展示眼镜，形成一个极强的视觉焦点。

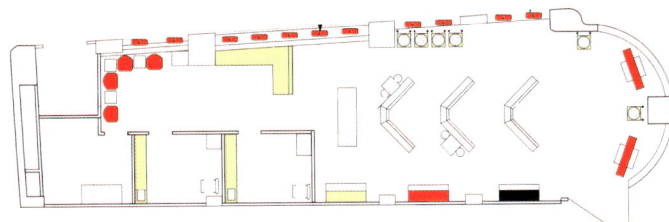

theeyecarecompa

FeiLiu Fine Jewellery

FeiLiu精品珠宝店

Project Name: FeiLiu Fine Jewellery   **Designer:** LISPACE   **Location:** Beijing, China
Photographer: LISPACE   **Time:** 2009

项目名称：FeiLiu精品珠宝店 设计师：北京立和空间设计事务所 项目地点：中国，北京 摄影：北京立和空间设计事务所
完成时间：2009年

FeiLiu Fine Jewellery is a British high-end jewelry brand; recently opened in Beijing Blue Harbor Solana shop is the brand's first exclusive jewelry store in Asia, offering customised services. FeiLiu Fine Jewellery's jewelry design has a strong sense of space, and the designers fully extended such features to the store's interior design.

The initial orientation of the store design is to break the traditional way, to introduce the concept of Gallery experience. The main space is a showcase of different sizes, highlighting both the product and a sense of display sequence. Interior colours reflect a neutral fashion. Dark brown is adopted, and the brand's main colour purple is used for the background colour and the logo. Interior lighting design does not emphasise decorative function; they remain focused on products. Some showcases are equipped with luminous lamps, with auxiliary lighting throughout the space. The space of the store, the overall design and the lighting contribute to create the gallery atmosphere, so that guests, when purchasing the jewelry, would at the same time experience the "Jewelry Museum" atmosphere.

FeiLiu Fine Jewellery is small and exquisite. The cleverly divided display areas include precious metals, silver display area, cuffs display area, showing wear areas, VIP custom room, dressing rooms, and a rational allocation of office space of 150 square metres.

The jewelry store design, in a limited space to meet the functional needs of the space's exaggerated shape and style of the brand's jewelry design, is a successful experiment for a commercial space design.

FeiLiu是英国的一个高档珠宝首饰品牌，新近在北京开的这家FeiLiu精品珠宝店是其在亚洲的第一家提供定制服务的专卖店。FeiLiu精品珠宝设计的首饰有很强的空间感，于是设计师将这一特点在专卖店的室内设计中加以延伸。

设计开始定下的理念是打破传统，引进美术馆的体验。店内的主要空间摆放了各种规格的展示柜，既突出了珠宝首饰，又强调了展示的有序性。室内的色彩以时尚的中性色——棕色为主，配合背景和品牌标识所用的紫色。室内照明设计并不强调装饰功能，而是以展示的产品为焦点。有些柜台配备了台灯，整个空间另有辅助照明。店内的空间感、整体的设计以及灯光共同营造了一种美术馆的氛围，顾客在选购首饰的时候同时能体验到一种"珠宝博物馆"的感受。

FeiLiu精品珠宝店不大但很精致。设计师聪明地将展示区分开来，包括贵重金属区、银饰区、腕饰区、佩戴展示区、贵宾特别区、更衣室以及一个150平方米的办公空间。

这家珠宝店的设计，在一个有限的空间内，满足了空间夸张的形态和珠宝品牌风格的双重要求，可以说是商业空间设计的一次成功的尝试。

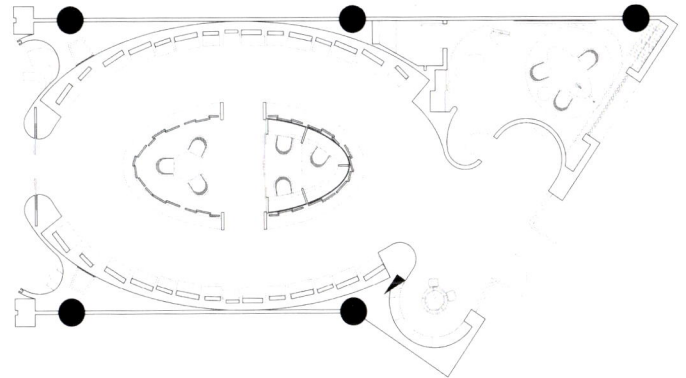

Couronne

钟表珠宝店

Project Name: Couronne   **Designer:** GLAMOROUS co.,ltd   **Location:** Tokyo, Japan
Photographer: Nacasa & Partners, Inc.   **Time:** 2007

项目名称：钟表珠宝店 设计师：美好设计有限公司 项目地点：日本，东京 摄影：姚京摄影 完成时间：2007年

The shop is located on the ground floor of a building designed by Tadao Ando in a peaceful quiet neighborhood and they sell very selected iconic watches from antique to cutting edge of the world. When you come in the shop, you will be overwhelmed by the beautiful copper colouration such as numerous wood beads hanging from the high ceiling.

The elegant showcases are surrounded with warm atmosphere from the natural materials such as wood beads and flooring, which successfully flatters the gorgeous watch selection. The central column has mirrors to broaden the space visually and enhance visual attractions. The luxurious continuous chandeliers standing on the copper-colour carpet also enhances the beauty of brilliant watches. The VIP room has artworks collected by the owner and classy furniture like an exclusive art gallery.

商店选址在一个安静平和的街区内一幢建筑的一楼（整幢建筑由安腾忠雄操刀设计），主要出售精选钟表商品，风格涵盖古典和时尚。走进商店，会情不自禁地被美丽的黄铜色调而吸引，木头的珠子从高高的屋顶上垂落下来，特色十足。

精美的陈列橱窗被温馨融合的氛围环绕，从而使得上面的华丽钟表更加突出。中央廊柱上镶嵌着镜子，在视觉上扩大了空间的面积；奢华的枝形装饰灯矗立在古铜色的地毯上，更加映衬出钟表的魅力；VIP房间内展示着店主收藏的艺术品以及优质的家具，俨然一个无与伦比的美术馆。

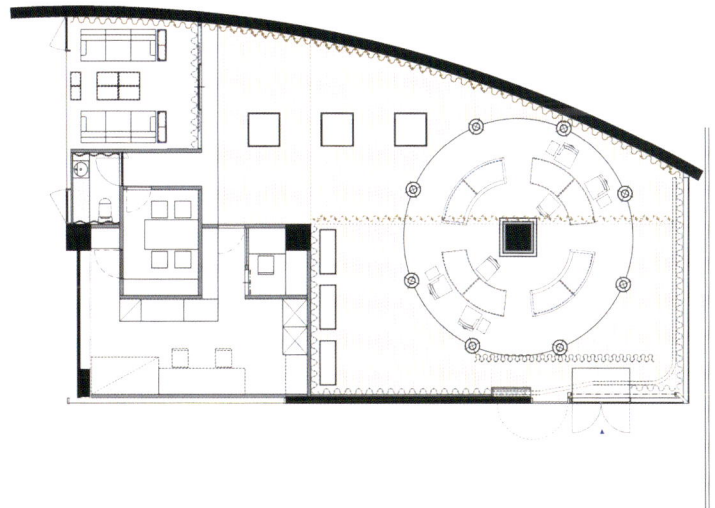

VAID Ginza

VAID珠宝专卖店

Project Name: VAID Ginza   **Designer:** GLAMOROUS co.,ltd   **Location:** Tokyo, Japan
Photographer: Nacasa & Partners, Inc   **Time:** March, 2007

项目名称：VAID珠宝专卖店　设计师：美好设计有限公司　项目地点：日本，东京　摄影：姚京摄影　完成时间：2007年3月

Italy has a traditional craftsmanship of jewelry working. VAID, one of the best jewelry brand for its famous handmade chain, was established in 1929 in Rome, Italy. Their hand-rolled and hand-knitted chains with fine subtlety and delicacy are especially acclaimed by famous jewelry shops throughout the world.

GLAMOROUS received the offer from Samantha Thavasa Japan Limited to deign the shop with respects for such a traditional brand image. The principal designer, Yasumichi Morita, designed the shop as it would lead the branding image of "VAID". Based on the concept of "glamour and luxury", he produced the glamorous space, hanging the copper brown chains in wavy pattern with the lighting effect to introduce luxurious VAID's famous "handmade chains".

It is not easy to create the active air into the shop with the solo specific material; however, he took advantage of the texture of copper brown chains, giving theme a feeling of floating with lighting effect so that all chains look like dancing in the air in luxury. He believes that these dramatic chains will provide every customer great anticipations of high jewelries to be seen.

意大利拥有传统的珠宝手工制作工艺。VAID作为著名珠宝品牌之一成立于1929年，其手工制作的链子因其精致的工艺和美感备受世界各地珠宝店的青睐。

美好设计受Samantha Thavasa（日本时尚品牌）公司之托为其设计VAID珠宝店，要求尊重品牌的古老形象。设计总监森田恭通亲自操刀，注重突显VAID品牌形象。以"美丽奢华"为中心理念，森田打造了魅力十足的空间，棕色的链子装饰形成波浪的形状，在灯光的照射下，使得VAID手工项链成为绝对的主角。

用同一种材料在店内营造欢快活跃的氛围并不是容易的事，但森田充分利用棕色链子结构的纹理，制造漂浮感。在灯光的照射下，一条条的链子如同跳舞一般，吸引着顾客的眼球。

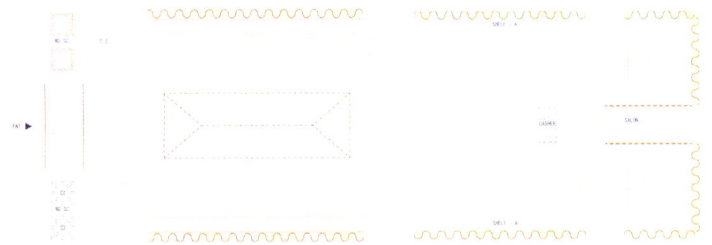

Index 索引

OFFICE

Project: Paul, Hastings, Janofsky & Walker, LLP – Paris
Design: Progetto CMR – Massimo Roj Architects

Project: Paul, Hastings, Janofsky & Walker, LLP – Los Angeles
Design: Progetto CMR – Massimo Roj Architects

Project: Manchester Square
Design: SHH

Project: Jamba Juice Support Centre
Design: Pollack Architecture

Project: Gummo Advertising Agency
Design: i29 interior architects

Project: Larchmont Office
Design: Rios Clementi Hale Studios

Project: New Google Meeting Room
Design: Camenzind Evolution Ltd.

Project: Baroda Ventures
Design: Rios Clementi Hale Studios

Project: Office Herengracht 433, All Capital
Design: i29 interior architects /
eckhardt&leeuwenstein architects

Project: Saatchi + Saatchi LA
Design: Shubin + Donaldson Architects

Project: Office Lab
Design: architects lab

Project: Turner Duckworth Offices
Design: Jensen Architects

Project: The Information Box
Design: neri & hu design and research office

Project: FOX Latin America Channel base offices
Design: Alberto Varas & Asociados, arquitectos
Estudio Angélica Campi

Project: China Everbright Limited (CEL)
Design: Lam Cham Yuen & Lee Ming Yan, Olivia

Project: Ernst & Young University
Design: YO DESIGN LIMITED

Project: An International Firm – Shenzhen Branch
Design: Lam Cham Yuen & Lee Ming Yan, Olivia

Project: Chengdu Sales Office
Design: One Plus Partnership Limited

Project: Machiya Office
Design: Hayakawa Hua Shi

Project: City Year Headquarters for Idealism1
Design: Elkus manfredi Architects

Project: Çağdaş Holding Office
Design: Nagehan Acimuz

Project: Meinl Bank
Design: ISA STEIN Studio

Project: DITTEL – Architekten
Design: d-arch

Project: Headquarters Expansion
Design: KlingStubbins

Project: The LG Air-conditioner Academy
Design: Illés Attila, Sotkó Anikó

Project: Ding Pu Hi-tech Square
Design: Lin Eddy E+ interior design studio

Project: TIC
Design: Noriyuki Otsuka Design Office Inc.

RESTAURANT

Project: Shiyuan restaurant
Design: Lispace design studio,LiJia

Project: Maedaya Bar
Design: Architects EAT

Project: Adour Alain Ducasse
Design: Rockwell group

Project: Adour at The St. Regis Washington, D.C.
Design: Rockwell group

Project: Dos Caminos
Design: Rockwell group

Project: Matsuhisa
Design: Rockwell group

Project: Wildwood Barbeque
Design: Rockwell group

Project: Tiandi Yijia – Restaurant
Design: Mauro Lipparini

Project: BEI – Asian Restaurant
Design: Lyndon Neri and Rossana Hu

Project: Sureno (Mediterranean Restaurant)
Design: Lyndon Neri and Rossana Hu

Project: Cityscape Restaurant
Design: Matt Gibson Architecture+Design Studio

Project: NEVY
Design: Concrete Architectural Associates

Project: Vengeplus Capacity
Design: Nagehan Acimuz

Project: SEVVA Restaurant
Design: TsAO & McKOWN Architects

Project: BRAND Steakhouse
Design: GRAFT

Project: Inamo
Design: Blacksheep

Project: The Levante Parliament
Design: Michael Stepanek

Project: "pearls & caviar"
Design: concrete

Project: Lido
Design: JP Concept Pte Ltd

Project: VLET Restaurant
Design: JOI-Design GmbH, Hamburg

Project: Chocolate Soup Café
Design: pericles liatsos designers

Project: Mc Donalds Urban Living Prototype
Design: Studio Gaia, Inc

Project: Oth Sombath Restaurant
Design: Jouin Manku

Project: Restaurant Alain Ducasse
Design: Patrick Jouin

Project: Leggenda Ice Cream and Yogurt
Design: SO Architecture

Project: Rosso Restaurant
Design: SO Architecture

Project: VYTA – Boulangerie
Design: Colli+Galliano Architetti

Project: EL JAPONEZ
Design: Cheremserran

Project: Charcoal BBQ 3692
Design: studiovase

Project: Le Square Restaurant
Design: MOBIL M

Project: Sky 21
Design: Danny Cheng Interiors Ltd

HOTEL

Project: Kush 222
Design: Johnny Wong and Miho Hirabayashi

Project: Chambers MN
Design: Rockwell group

Project: Carbon Hotel
Design: PCP Architecture

Project: Klaus K Hotel
Design: SARC Stylt

Project: The Dominican Hotel
Design: Lens Ass Architects

Project: Jeronimos Hotel
Design: Capinha Lopes& Associates

Project: CitizenM Hotel
Design: concrete

Project: Murmuri Hotel
Design: Kelly Hoppen

Project: Hotel Sezz
Design: Shahé Kalaidjian, Christophe Pillet

Project: The George Hotel
Design: synergy hamburg

Project: Villa Florence Hotel
Design: sfa design

Project: JW Marriott Hotel Hong Kong
Design: JW Marriott Hotel Hong Kong

Project: Hotel Pirámides Narvarte
Design: DIN interiorismo, Aurelio Vázquez Durán

Project: Mövenpick Airport Hotel
Design: Matteo Thun & Partners

Project: The Library
Design: Tirawan Songsawat

Project: X2 Koh Samui
Design: Be Gray Limited

CULTURE

Project: Thurles Arts Centre and Library
Design: McCullough Mulvin Architects

Project: A.E. Smith High School Library
Design: Atelier Pagnamenta Torriani

Project: Wagner Middle School Library
Design: Atelier Pagnamenta Torriani

Project: Ps 11r Primary School Library
Design: Atelier Pagnamenta Torriani

Project: The Danish Jewish Museum
Design: Tomrerfirma Gert Fort A/S

Project: Museum
Design: JSª

Project: New Acropolis Museum
Design: Bernard Tschumi Architects

Project: The Art Institute of Chicago – The Modern
Wing, Chicago
Design: Renzo Piano Building Workshop, Paris,
France

Project: The Arts of Asia Gallery, Auckland
Design: Catherine Stormont, Nicole Pfoser, Gustavo
Thiermann

Project: That's Opera
Design: Atelier Brueckner GmbH

Project: Bachhaus Eisenach
Design: Penkhues Architekten, Kassel

Project: Multikino Szczecin
Design: Robert Majkut Design studio

Project: Blue Room Theatre at Chesapeake
Design: Elliott+Associates Architects

Project: Ordrup School
Design: Bosch & Fjord

Project: Medical Faculty, University of Groningen
Design: RAU

FITNESS CENTRE

Project: Chesapeake Fitness Centre, East Addition
Design: Elliott + Associates Architects

Project: 14 Street Y Renovation
Design: Studio ST Architects

Project: Y+ Yoga Centre
Design: Lyndon Neri and Rossana Hu

EXHIBITION

Project: Breaking New Ground
Design: Jutta Friedrichs, Ben Houge

Project: Butterflies + Plants
Design: Reich + Petch Design International

Project: Fuji Xerox Epicenters
Design: Geyer Pty Ltd

Project: Stylecraft Showroom
Design: Geyer Pty Ltd

Project: Bernhardt Design Showroom
Design: Lauren Rottet, Kelie Mayfield, Christopher
Evans

Project: BRANCH in Changchun
Design: Keiichiro SAKO, Yoshimasa TSUTSUMI /
SAKO Architects

Project: FlatFlat in Harajuku
Design: Keiichiro SAKO, Yuichiro IMAFUKU / SAKO
Architects

Project: homeculture
Design: Franken Architekten

Project: GRAFTWORLD – Exhibition in the Aedes
Gallery
Design: Graftlab

Project: Moonraker
Design: Lam Cham Yuen & Lee Ming Yan, Olivia

Project: Internorm Flagshipstore
Design: ISA STEIN Studio

Project: Heaven
Design: Capella Garcia Arquitectura

STORE

Project: Tesla Store
Design: CCS Arhictecture

Project: ProNature
Design: Joey Ho / Joey Ho Design Ltd

Project: Healthy Life Natural Health Food Store
Design: Design Clarity

Project: Barbie Shanghai
Design: Slade Architecture

Project: WEEKDAY MALMÖ
Design: Electric Dreams

Project: Monki Stockholm
Design: Electric Dreams

Project: MONKI 1: Forgotten Forest
Design: Electric Dreams

Project: LEVI'S BB Barcelona Summer
Design: Jump Studio

Project: Forum Duisburg
Design: Chapman Taylor Architects

Project: Sexta Avenida
Design: Chapman Taylor Architects

Project: Celebrity Solstice
Design: 5+design project team

Project: Gourmet
Design: HEAD Architecture and Design Limited

Project: Only Glass
Design: Robert Majkut Design studio

Project: Rocawear Mobile RocPopShop
Design: d-ash design

Project: SENSORA
Design: Design Clarity

Project: UHA Mikakuto
Design: GLAMOROUS co., ltd

Project: O2
Design: JHP

Project: Lefel–The Whispering World
Design: Crea International

Project: Nike Airmax 360 Exhibition
Design: Mark Lintott, Chia Yen, Alison Yang

Project: Kymyka shoes and bags, Maastricht
Design: Maurice Mentjens Design, Holtum

Project: Offspring Camden
Design: Shaun Fernandes, Sarah Williams

Project: Sabateria Sant Josep
Design: Joan Casals Pañella / CSLS arquitectes

Project: The Eyecare Company
Design: Design Clarity

Project: FeiLiu Fine Jewellery
Design: LISPACE

Project: Couronne
Design: GLAMOROUS co., ltd

Project: VAID Ginza
Design: GLAMOROUS co., ltd